MW00911196

NEW INFOTAINMENT TECHNOLOGIES IN THE HOME

Demand-Side Perspectives

LEA's COMMUNICATION SERIES
Jennings Bryant/Dolf Zillmann, General Editors

Selected titles include:

Clifford/Gunter/McAleer Television and Children: Program Evaluation, Comprehension, and Impact

Biocca/Levy Communication in the Age of Virtual Reality

Gershon The Transnational Media Corporation: Global Messages and Free Market Competition

Harris Health and the New Media: Technologies Transforming Personal and Public Health

Salvaggio The Information Society

For a complete list of other titles in LEA's Communication Series, please contact Lawrence Erlbaum Associates, Publishers

NEW INFOTAINMENT TECHNOLOGIES IN THE HOME
Demand-Side Perspectives

Edited by

Ruby Roy Dholakia
Norbert Mundorf
Nikhilesh Dholakia
RITIM,
University of Rhode Island

LEA LAWRENCE ERLBAUM ASSOCIATES, PUBLISHERS
1996 Mahwah, New Jersey

Lawrence Erlbaum Associates, Inc., Publishers
10 Industrial Avenue
Mahwah, New Jersey 07430-2262

Library of Congress Cataloging-in-Publication Data

New infotainment technologies in the home : demand-side perspectives / edited by Ruby Roy Dholakia, Norbert Mundorf, Nikhilesh Dholakia.
p. cm.
Includes bibliographical references and index.
ISBN 0-8058-1626-7 (alk. paper)
1. Computer industry—United States. 2. Television supplies industry—United States. 3. Telecommunication equipment industry—United States. 4. Household electronics industry—United states. 5. Multimedia systems—United states. 6. Information technology—United States. I. Dholakia, Ruby Roy. II. Mundorf, Norbert. III. Dholakia, Nikhilesh, 1947-
HD9696.C63U526352 1996
381'.45621381'0973—dc20 96-1373
CIP

Books published by Lawrence Erlbaum Associates are printed on acid-free paper, and their bindings are chosen for strength and durability.

Printed in the United States of America
10 9 8 7 6 5 4 3 2 1

Contents

Preface

Much has happened since we organized a colloquium on "New Information Technologies: Panacea or Peril?" at the University of Rhode Island during Spring 1993. Even as the players in the mega-merger game have changed—TCI and Bell Atlantic have now been replaced by Microsoft and NBC—the theme has remained unchanged, a convergence of the information and entertainment businesses to create a new infotainment industry. It is easy to predict this convergence but still difficult to determine how the new industry will develop and deliver goods and services that will be accepted by masses of consumers.

One development that seems to have eclipsed the unfolding of the infotainment industry is the phenomenal growth of the Internet and World Wide Web. The access to this global network by millions of users and businesses will fundamentally alter the ways in which we design, deliver, and consume infotainment goods and services. This book does not directly address issues regarding the Internet and the World Wide Web, but many of the conceptual and analytical frameworks presented here are clearly applicable broadly to all information technologies.

This book reflects the thinking on the supply- and demand-side factors that will shape the infotainment industry. Because of the nature of the topic, several details change daily but the basic issues retain their importance and centrality. The complexity of technologies involved in designing and delivering established and new services means that a technology push from the supply side will continue to extend the frontiers of the infotainment landscape. New players will emerge and established players may retire, exit, or perish, but the supply side will continue to exert its pressure relentlessly.

Demand-side factors will determine whether the supply-side pressures will be received with welcome arms and open pocket books or met with yawning indifference and even direct hostility from the consumer. The macro environment, including public policy, will also play a big role in the ways supply and demand-side factors converge. Public policy, however, is not created in a vacuum and the demand- and supply-side factors will contribute to its shape and force. This collection provides an insightful and diverse set of perspectives, especially from the demand and public policy sides, that help us understand the unfolding shape and character of the age of infotainment.

We would like to thank the authors who contributed to this volume and worked over many months to write, revise, and edit their chapters. Because of a mishap at the printing shop, they were all forced to essentially repeat their efforts and for their prompt response, we are most grateful. We are indebted to Dr. Jennings Bryant, Director of ICR, University of Alabama and series editor at Lawrence Erlbaum Associates Publishers for having encouraged us to develop the proposal for this book. Robin Weisberg at Lawrence Erlbaum Associates worked with us to bring this book to publication.

Financial support for the colloquium was provided by the Honors Program at the University of Rhode Island as well as by NYNEX. Mary Ann Shope was our champion within NYNEX and helped us with many arrangements. We are particularly grateful to Carol Grant, Vice President of NYNEX, Rhode Island for her continued support. The Research Institute for Telecommunications and Information Marketing (RITIM), at the College of Business Administration, University of Rhode Island, provided support in terms of graduate assistance and financing various aspects of this project.

—RRD, NM, and ND

1
Bringing Infotainment Home: Challenges and Choices

Ruby Roy Dholakia
Norbert Mundorf
Nikhilesh Dholakia
The University of Rhode Island

Most experts agree that the future of information and entertainment will be the result of a merger of several technology sectors, including computers, telephone, broadcast, and cable television. Until recently, these sectors remained clearly separated both technologically and organizationally. The AT&T divestiture and Judge Greene's Modified Final Judgment (MFJ) kept the phone and cable industries clearly separated. Cross-ownership rules and other regulations limited the involvement of broadcasters with other technology sectors.

Major changes are afoot in the 1990s. The prospect of a mass market 500-channel cable system, which can also transmit voice and computer data, has come within tangible reach. Major contributing technical factors are improvements in digital technology, notably compression, which permits carriage of multiple video signals on one conventional cable channel, and the widespread implementation of optical fiber in the "backbone" of cable and telephone networks. These developments coincided with significant alliances involving the cable, telephone, film, and broadcast industries and announcements of major upgrades of the cable and telephone infrastructure, as well as agreements on high definition television (HDTV; Fabrikant, 1993; "The Future of TV," 1993; "A Phone-Cable Vehicle," 1993).

There have been a number of mergers and alliances crossing traditional telecommunications industry lines (N. Dholakia & Dholakia, 1995). Most of the seven Regional Bell Operating Companies (RBOCs) in the United States are involved in telecommunications mergers and acquisitions ("A Year of Turmoil," 1993), although some of the more visible ones have been cancelled. Software giant Microsoft has initiated a process of mergers and partnerships to transform itself into the premier communications services firm on the information highway (Gilder, 1994; Schlender, 1994).

Some RBOCs are striking an independent path. Pacific Bell recently announced a $16 billion investment in its telecommunications infrastructure. In contrast to other arrangements, this one does not include cable companies. Pacific Bell is skeptical regarding further deregulation and is focused on competing against rather than collaborating with the cable industry. SNET and Ameritech are also following independent strategies. For example, Ameritech was granted permission to deploy commercial video-dialtone service (Hontz, 1995).

In Europe, the former British, French, and German Telecommunications monopolies are currently undergoing significant changes (Steinfield, Bauer, & Caby, 1994). They are also active in implementing broadband fiber optic technologies that will permit interactive services carrying voice, data, and video. Currently, the emphasis is on work applications, such as the communication between government agencies in the "old" capital of Bonn and the "new" capital of Berlin (Witte, 1993). To make a nationwide fiber optic broadband system financially feasible, however, home use by private consumers will be pivotal (Mundorf & Zoche, 1993).

APPROACHING INFORMATION TECHNOLOGY

Information technologies can be approached either from a supply-side perspective that focuses on the industries and firms that make and market information technology products and services, or they can be approached from a demand-side perspective that focuses on the users of information technologies. Users of information technology can be either in the business sector or they can be people at home. Combining these two dimensions gives us a four-quadrant framework for approaching and analyzing the applications and impacts of information technologies (see Fig. 1.1). There is great variability in the amount and type of knowledge that exists about the applications of information technologies in these four quadrants. The structure of knowledge in these quadrants is examined here.

In the first quadrant, the supply-side aspects of information technologies targeted at business users, the key issues deal with industrial structure and organization. Industry studies, primarily in the proprietary domain of the

	Business	Home
Supply-side	I	III
Demand-side	II	IV

FIG. 1.1. Four-quadrant framework.

consulting world, but also some in the publicly available literature provide many insights into the conditions and factors that govern the supply of information technologies to the business users (see, e.g., Byrne, 1993; R. Dholakia, 1994a; Malone & Rockart, 1991). The second quadrant has been the subject of some consulting studies, such as the periodic surveys of information technology use in the business world by Link Resources, and from the 1980s a few significant academic studies have examined how organizations use information technologies (see, e.g., Kiesler & Sproull, 1992; Zuboff, 1988).

There is a growing interest in the third quadrant—the supply of information technologies to households—from the popular media in the 1990s, especially following many mega-mergers and alliances in the cable, telecommunications, software, and computer industries aimed at tapping into the anticipated home market for a variety of interactive services ("25 Breakthroughs," 1994; Schlender, 1994). Some analytical insights on this quadrant can be found in the collection of papers edited by R. Dholakia (1994a).

THE NEGLECTED QUADRANT

Little systematic research exists on the demand-side of information technology in the home (i.e., the fourth quadrant of Fig. 1.1). Much of the writing on information technologies has been technology-driven, resulting in speculation about what technological changes will occur in the home. Alternatively, it has taken a regulatory or economic viewpoint from the perspective of one or the other industry (e.g., "The Future of TV," 1993). The prevailing mode has been what has been labeled a *Field of Dreams* approach: "If we build it, they will come." Implied in this "building" is the construction and promotion of interactive, broadband multimedia networks and devices for the home, suffused with all manners of software and services, although little is known about what consumers actually want. Findings from the recent advanced cable trial in Queens, New York, give some indications. More choice tends to increase sales; the ability to pause and resume a movie is a critical feature; younger people find it easier to deal with technology than older people. As far as the features of the system are concerned, added channel capacity in and of itself is insufficient. Instead, the novelty of the experience is critical. And finally, a good navigational system is of critical importance (Auletta, 1994). Judging from recent articles in the popular or trade press, the issue of bringing interactive, broadband, multimedia technologies to the home has caught fire, at least in terms of journalistic and public imagination (e.g., Blankenhorn, 1993; Elmer-Dewitt, 1993; "The Future of TV," 1993; "Media Mania," 1993; "A Year of Turmoil," 1993).

CHALLENGES

Although visionary dreams help fire up the imaginations of people—consumers, journalists, public policymakers, business leaders, and technology developers—specific decisions to invest money, time, and effort require a deeper, systematic, and research-based understanding of which information technologies are likely to be accepted in the home, how they are likely to be used, and what impact are they likely to have on the lives of people. In the path of such understanding lie many challenges, to which we now turn our attention.

Content and Selection

Although the technological development and implementation of the 500-channel or more TV systems seems to be just a matter of time, its content and usage is subject to conjecture by experts from industry, politics, and academia (cf. N. Dholakia & Mundorf, 1993; Haugsted, 1993). These speculations are based on current market trends, and probably a great degree of "hype" because most Americans are unsure about the future use of information technologies. In a recent survey, 59% of respondents reported that they understand the concept of the information superhighway "not very well or not at all" (Maddox, 1994). The prevailing impression seems to be, however, that it has to do with business (37%) rather than entertainment (8%) or science (5%).

Many corporations involved in the "construction" of the superhighway are currently betting on entertainment as the engine ("The Entertainment Economy," 1994). This view extends beyond the large cable, phone, and video enterprises to thousands of small businesses (Rebello & Eng, 1994), based on the seemingly insatiable appetite of Americans for entertainment. Notwithstanding recession and unemployment, household expenditures for entertainment and recreation have risen steadily, from 7.71% of nonmedical consumer spending to 9.43% ("The Entertainment Economy," 1994). Assuming that most entertainment dollars are spent on media, this increase seems at odds with Carey's (1992) and Wood's (1986) assertion of a relatively constant proportion of media expenditures.

A small group of researchers has long asserted that entertainment is the key function—and effect—of media consumption (Zillmann, 1993). And some suggest that the main function of new information technology is to "pack fun into every nook and cranny of the day" ("The Entertainment Economy," 1994, p. 64). Although much of the growth of recent years has been in gambling and theme parks, it is conceivable that both of these types of entertainment will soon be accessible from the home, through video gambling and games involving virtual reality.

Although some existing theories may be pertinent to future media choice (e.g., theories of selective exposure and mood management; Westin, Mundorf, & Dholakia, 1993; Zillmann, 1988) or diffusion theory (Rogers, 1986), little empirical work exists that is directly pertinent to media choice in the 500-channel environment. Some researchers have discussed the role of commercial broadcasters in such an environment (see, e.g., Baldwin & Litman, 1992). They have focused on the industry rather than the audience side, however. Due to the rapidity of these developments, much of the writing has been in the popular and trade press, but little academic research has been conducted.

Many writers have expressed concerns about the possible lack of demand for the kind of variety provided by the 500-channel system. More importantly, they have called into question the willingness of consumers to pay significantly more for new services ("Media Mania," 1993). Part of this willingness will depend on the perceived utility of new content options. Given the relative constancy of budgeting for communication services (Carey, 1992; Wood, 1986), new services have to offer added value, ideally substituting expenditures from other parts of the household budget.

Interface and Usability

One of the uncertainties surrounding the new information technologies for the home is the kind of terminal device that will be feasible and popular with consumers. Television, with 98% penetration into U.S. homes, is the obvious candidate, followed by the telephone, with 97% penetration. Computers are much more versatile as information processors than present-day televisions, but their penetration rate into the home sector crossed 30% only in the early 1990s (see Table 1.1 for the ownership levels by income categories of personal computers and other recent electronic products). This, of course, is in the United States (R. Dholakia, 1993) and in most

TABLE 1.1
Househhold Income and Ownership of Selected Electronic Products in United States (in percent)

	Income Categories					
	Less than $20,000	*$20,000–$29,000*	*$30,000–$39,000*	*$40,000–$49,000*	*$50,000–$99,000*	*More than $100,000*
Answering machine	26%	37%	48%	48%	59%	60.8%
Personal computer	9%	19.1%	24%	37%	43%	54%
Fax machine	1%	1%	1%	4%	8%	17.6%
VCR	61%	84%	90%	94%	95%	94.1%
Video-game console	19%	36%	36%	41%	44%	35.3%
CD player	14%	27%	30%	39%	47%	52.9%

Source. Link Resources (1992).
(Sample = 2,500 in United States.)

other countries the penetration rate of computers is lower. Of the computers in the home, only about one fourth are equipped with modems. And of these computers, a very small portion have dedicated telecommunications lines available for receiving 24-hour interactive programming. These represent major barriers to the use of computers as the main terminal device for the multimedia interactive services of the future. By contrast, more than 70% of U.S. homes are passed by cable and more than 60% are cable TV subscribers. It is no wonder that a CATV/phone combination is being seen by the industry as the easiest way to bring these new services into the home, at least at the initial stage (Lieberman, 1994). As TV sets get "smarter," they will acquire the functionalities of a computer, and the versatility advantage of computers will disappear. In fact, as television sets get smarter and computers acquire audiovisual capabilities, the issue will no longer be framed in "computer versus TV" terms, but in terms of the type of network to which these devices are connected. Recognizing this, Microsoft's strategy is to create a ubiquitous software-driven network capable of serving any smart device that uses a Microsoft interface. Homes and businesses served by such a network will be able to interact with various forms of "infotainment" in versatile and friendly ways. Given Microsoft's past success in dominating markets, this new strategy worries those building competing networks but reassures thousands of device makers and content providers that a fairly ubiquitous and usable interface will emerge.

The discussion on interface devices here, although of value to industry decision makers, is still framed in supply-side terms: It is concerned with which types of devices and networks can be made available readily to most homes. From the demand side, the issue has to be framed differently: How would consumer want to interact with interactive multimedia networks? This issue has been examined in myriad ways, including considerable amount of "user-factors" research and substantial work on the "Remote Control" as a preferred interactive device (Walker & Bellamy, 1993). The overall lesson from such research is that using the remote control modifies the viewing experience by enabling the linking of seemingly unrelated content into a new whole. Remote controls have also modified the behaviors of advertisers and content programmers. At a more basic level, the issue should be framed in terms of consumer culture: How do people see new technologies and services shaping their life? Are some patterns of interaction with technology more consistent with the way consumer culture and nonwork life are evolving than others?

Economic Issues

Americans spend about 50% of their household budgets on housing and transportation, whereas media and telecommunications take a relatively small share (Carey, 1992). In addition to shifting priorities within the media and telecommunications budgets (e.g., from video rental to pay-per-view [PPV]), significant increases will have to come from other areas, such as housing and transportation. By minimizing the need for daily commutes to the workplace, telecommunications could make living in less costly (and more appealing) areas more feasible. Arthur D.Little, Inc. has estimated savings of $23 billion (in 1988 dollars) per year if 10% to 20% of U.S. road trips were replaced by telecommuting (Eckersman, 1991). Authors of one survey estimated that telecommuters saved an average of $2,000 (in 1990 dollars) a year by not driving to work, eating out at noon, and buying work clothes (Roderick & Jelley, 1990). This trend is currently noticeable in a fairly small group of "lone eagles" ("All Things Considered," 1994). But the reduced cost and size of equipment, improvements in transmission capabilities, lifestyle, and greater flexibility of the workforce could broaden this trend.

Time and travel expense may be saved through home shopping. Such benefits may be particularly appealing to two-income households with young children, whereas others may see the shopping experience as an enjoyable social event (cf. R. Dholakia, 1994c; R. Dholakia, Pedersen, & Hikmet, 1995; Meyer & Schulze, 1993). The more vivid (Brosius & Mundorf, 1991) and interactive the home shopping experience becomes, the greater the appeal within and beyond this core group (Mundorf, Dholakia, Dholakia, & Westin, 1995; Mundorf, Westin, & Dholakia, 1993). For instance, elderly consumers may find shopping tedious, but won't give it up because of the social component. If home shopping provides the opportunity to interact with salespeople, and maybe even other shoppers electronically, it may provide a safe and convenient alternative (Saloman & Koppelman, 1988).

Such substitution may also take the form of "vicarious" adventures and experiences. In this case, the technology provides a service with a cost–benefit ratio better than "the real thing." Travel may be an example. As Saffo (1993) pointed out, however, increased telecommunications did not diminish but actually increased travel to keep up with the greater number of contacts that were established. There seems to be some evidence to the contrary also. Garhamner and Gross (1993) found that telecommuters do indeed travel less than their colleagues with conventional work arrangements.

Life-Space Issues

In the industrial era, people's life-spaces were demarcated into the categories of work and personal life. With increasing complexity of life, people's personal lives were further divided into activities in the home such as cooking, cleaning, and watching television and activities outside the home such as shopping, recreational outings, and banking. Traditonal information technologies were clearly separated as far as their functions in society were concerned: radio, TV and stereos were used for private entertainment in the home; computers, robots and other automated equipment was clearly work-related; and there was a third set of activities and services related to the transactional sphere between work and play—shopping, banking, insurance, travel, taxes, voting, and so on. New information technologies create the possibilities of re-engaging and perhaps even integrating the three spheres of people's lifespace: work, homelife, and transactions (see Fig. 1.2).

A mainstay in the workplace, computers are being used more and more for transactional functions and private household activities (R. Dholakia, Dholakia, & Pedersen, 1994). Pressures to enmesh the home and the work spheres come from social concerns regarding air quality, energy costs, real estate, and labor costs as well as organizational forces (Sell & Jacobs, 1994). The success of telecommute programs are determined as much by these supply-side characteristics as demand characteristics such as individual motivations and demographics. Similarly, for transactional activities such as teleshopping, telebanking, and telemedicine, supply-side considerations of the type mentioned are as important as demand-side considerations.

Once a technological system gets established in one life sphere, however, it has the potential to migrate to the other connected spheres. In Japan, for

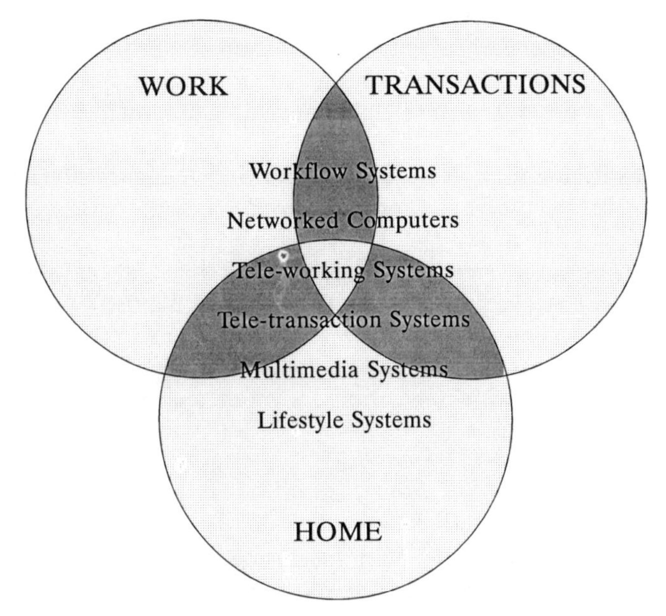

FIG. 1.2. Integration of life spheres and technologies.

example, Nintendo re-positioned game machines as "Famicon," a device that could be used to play games as well as do activities like banking.

Lifestyle and Family Interactions

New information technologies are often presented as technologies of connection that will bring families and friends closer, regardless of physical distances, or time schedules (Christen, 1987; Young, 1992). On the other hand, these same technologies have the potential of cocooning individuals into their own virtual worlds and speeding up the ongoing rush into private, passive, and alienated lifestyles (Fırat & Dholakia, 1982). In which ways will new technologies shape family and community lives? The answer to this questions will have a major bearing on how these technologies are received by society.

Flexible use of time, better scheduling, time savings, and improved accessibility to children are all seen as advantages of teleworking. On the down side, some miss a clear separation of private life and work, whereas others report intensified family contacts (Mundorf, Meyer, Schulze, & Zoche, 1994). Gender seems to moderate the effect of telecommute programs on work and nonwork life (Sell & Jacobs, 1994); women telecommuters for example, reported less leisure time, more work time, and more family conflict over the use of personal space and time for work ("Still on the Line," 1985).

CHOICES AND CONSEQUENCES

Choices and Behaviors

As the supply-side environment evolves through the development of technology and industry restructuring, the user at home will be faced with a changing spectrum of choices. Entrenched patterns of electronic gadget ownership and telecommunication (including cable TV) subscriptions, however, will constrain the user from changing too fast and too far. It is unlikely, for instance, that the VCR will be thrown out soon to watch movies on demand even though watching movies is the primary use of the VCR (see R. Dholakia, this volume; Mundorf & Dholakia, 1994). As Orme (1994) reflected on the results from video-on-demand trials in Littleton, Colorado, consumer usage is likely to be greater than the existing pay per view, although it will not yet be a great threat to traditional video stores. What we are likely to witness in the years to come is a long battle of information delivery formats in the home.

From our understanding of consumer behavior with respect to information products, services, and media, we can draw some general conclusions about the types of choices users at home will make. Consumers are likely to choose tangible product choices over intangible service offers (R. Dholakia 1992, 1994b; R. Dholakia & Venkatraman, 1993); consumers are likely to invest in peripheral devices that provide entertainment to multiple members of the household (R. Dholakia, 1993) as well as those that facilitate polychronic time use (Kaufman & Lane, 1993), especially by time-harried consumers.

Resources and Constraints

Acquisition and usage of information technologies, products, and services by home users is going to be constrained by the availability of four major categories of resources to the user: money, time, space, and skill. These resources are unevenly distributed in the U.S. population.

Money Resource. Data on penetration of various entertainment and information technology products show that the adoption of newer technologies increases dramatically with the average annual income. Additional research on pricing of new information technology products suggests that the average price for a new product has been falling while achieving lower levels of penetration (Carey, 1993). The combination of higher income and lower price thresholds creates significant barriers for the adoption and diffusion of new products. Income is no longer an increasing resource among U.S. households. Between 1980 and 1990, median earnings for year-round full-time workers actually declined for men and rose only slightly for women (Waldrop & Exter, 1991).

Time Resource. According to many researchers, time is of greater value than even money to specific market segments (Hornik & Schlinger, 1981). Time affects the adoption and utilization of technology in many ways: It may stimulate the choice of a specific technology, it may affect the spatial location of a technology, and it may impact the ways in which other resources are used (Kaufman & Lane, 1993). Kaufman, Lane, and Lindquist (1991) argued that polychronic or multiple time use is the preferred strategy for many consumers.

Sometimes a technology is monochronic in its involvement (rather than polychronic). Venkatesh and Vitalari (1987) reported time trade-offs as a result of adopting and using computers at home. In the early periods of use, computers lead to lower time spent on other activities, especially activities that are similar such as watching TV. The AT&T trials suggest that youths talked less on the phone when they used interactive TV (Keller, 1993).

Skill Resource. A certain level of skill has become a prerequisite for the use of many new household technologies even if "smart" technologies reduce the level of required skills. Some researchers have argued for "visual literacy" because of the dominant influence of the TV paradigm (Fırat, this volume).

The availability of skills and the motivation to utilize available skills vary. Reese (1990), for instance, found that clerical/service workers reported very low usage of the computer at home despite heavy use of the computers at work. Venkatesh and Vitalari (1987) reported a higher intention to use the computer for business and education purposes than actual use. Gender differences in the use of the computer have been the focus of much research (cf. R. Dholakia et al., 1994). The motivation to acquire and learn requisite skills also vary. In the trial of multimedia services, AT&T found that "eight year olds were teaching their parents how to use the services" (Keller, 1993, p. B1).

Space Resource. As technological devices and appliances fill U.S. homes, questions arise regarding the adoption and location of new ones. Presence of "wire clutter" is common. Despite or perhaps because of the larger sizes of the average U.S. home, space has not been a consideration in the adoption and diffusion of consumer products. In fact, it has facilitated the ownership of multiple units of a product such as telephones and TV sets.

Space, however, is becoming an important constraint on consumer behavior. Data on cable TV penetration in the United States suggests that although most households own multiple TV sets, the cable connection is likely for only one set, primarily located in the most public space such as the living or family room ("A Comparative Analysis," 1991). Recent research on HDTV suggests similar spatial constraints. Although a large-screen TV is best capable of demonstrating the benefits of HDTV, adoption potential appears to be limited because of consumer perceptions regarding spatial requirements of HDTV (Dupagne & Agostino, 1991).

Consequences and Learning

For many information products and services, the initial act of adoption is just the beginning of consumer learning. With time and usage, many things happen, including new applications, upgrading or downgrading of systems and services, idiosyncratic and novel adaptations (which often become the bases of future supply-side innovations), and migration to related infotainment products and services. Not only are there transformations in the things adopted, but the adopted things often transform people's lives in major ways—leading to further consumer behaviors that affect markets for information technologies, products, and services. Understanding the impact of technologies on people's lives will be crucial for marketers, policymakers, and consumer advocates.

PLAN OF THIS BOOK

Although it is impossible to cover in one volume all the challenges to a systematic understanding of information technology acceptance, adaptation, usage, and impacts, the writers in this collection have done an admirable job of providing tools and concepts to chip away at these challenges. Contributions in this book are organized into three categories: managerial perspectives, user perspectives, and policy perspectives.

Managerial Perspectives

In this section, the managerial perspectives on providing new infotainment services are examined through four chapters. In Chapter 2, "Market Opportunities and Pitfalls for New Information Technology in the Home," Samuelsson reflects on the many promises of new technologies and compares them to their actual market acceptance. While commenting that marketing hype is necessary for fast and dramatic introduction of new technologies, and despite the technology push, he finds that market acceptance has frequently been gradual and very different from what has been forecasted. All the successes are characterized by what, he calls the "wedge application," a use found by the consumers that "cracks open the market and continues to drive the penetration." Recognizing that it is easier to predict the past, Samuelsson feels that it will be the combination of both supply- and demand-side drivers that will lead to "wedge applications which are the result of spurs of innovation that create excitement among users or based on the fact that a clear market need can suddenly be met because of progress in the technology current."

Carey and Elton (chap. 3) analyze diffusion and adoption of household technologies over time. They show that traditionally, technologies have followed the adoption curve and attained mass market status once pricing had come down to an affordable level. Initially, these technologies depended on lead users to finance the slow adoption phase. For many new technologies, reliance on the adoption curve causes problems; most new technologies are only one of several alternates that basically serve the same function for the consumer. As such, they probably will never reach a high level of adoption like, for instance, the VCR. In fact, chances are good that a technology will never "take off" or only reach the desired point on the adoption curve 5 years "late."

In chapter 4, Orme's specific focus is on strategies for broadband services. He outlines the supply-side pressures that are pushing phone, cable TV, and entertainment companies to offer new services to the residential market. Given the nature of current demand, he concludes that all these efforts are likely to lead to increased choices for the consumer in terms of video entertainment but that it is unlikely to increase the total revenues for the major players. Instead, he argues that to gain wider acceptance of these new services, strategies have to be directed at those market segments that will benefit from an economic payoff from the adoption of such services. Only a few of the applications with economic payoff to the consumer are entertainment oriented such as shopping and games/gambling. Market potential, he feels, will be greater in the work-at-home segments and these users will serve to crack open the general home market. He also recommends telemedicine and tele-education applications because market drivers in these segments are currently promoting greater use of broadband services.

The emphasis on providing video entertainment by cable TV, telephone, and other media companies reviewed by Orme is examined more closely in chapter 5, by R. R. Dholakia. Movies-on-demand has often been cited as the number one killer application for new broadband network services that has attracted the attention of leading telephone, cable TV, and entertainment companies. Despite the large amounts of money spent on movies by consumers, particularly for video entertainment at home, movies-on-demand are seen to have limited appeal and mostly to those who find the concept "fitting with their lifestyle" choices. These consumers are those who view movies regularly and have already invested considerably in video technologies. It is precisely because of these consumer characteristics that appealing to this consumer segment limits the design of the new service. The chapter also highlights the importance of supply-side characteristics because technical complexities of implementing a true movies-on-demand concept still remain a challenge that must be surmounted before widespread deployment of broadband services can occur.

User Perspectives

Private consumers have a variety of reasons for purchasing and using particular technologies. These reasons are often not obvious to outside observers or even to the users themselves. Because these technologies are fairly new to the consumer market, we know little about this product category. Part II of the book presents five chapters illustrating various user dimensions of information technologies. The chapters complement each other and shed some light on different facets of consumer behavior, thus making sense of the puzzle of user perceptions, attitudes, and behaviors regarding information technologies.

Bryant and Love (chap. 6) explore trends in new information technology as reflected by media coverage and survey research. Their analysis indicates limited consumer interest in advanced technologies, and growing skepticism among experts regarding the rapid diffusion of information technology innovations among private households. The authors also point to the critical role of entertainment in the adoption of technologies by the increasingly important home market. Demonstrating the inadequacy of conventional research, their chapter discusses new theoretical approaches necessary to reach a good understanding of user responses to interactive media. As a contribution to such a theoretical framework, the authors identify key elements related to the entertainment experience. In their conclusion, they call for an entertainment theory for the information age, which would lead to better and more comprehensive research.

Cermak (chap. 7) uses clustering and multidimensional scaling to analyze perceived mental relationships between different entertainment alternatives. This technique makes it possible to understand the relationship between new technologies and traditional activities for which they may be substitutes—assuming that initially, substitution will be the main function of these new technologies. Cermak also sheds some light on the relationships between various new technologies. Entertainment and information media, especially those that are solitary and sedentary, tend to be clustered around in a small part of the consumer's mental map. Subdimensions of such a map include electronic versus nonelectronic and practical outcome (hobbies) versus pure entertainment. Because of the dense clustering of existing solitary/sedentary entertainment options in a corner of the consumer's mental map, new technologies face stiff competition in this area. Marketers will have to come up with strategies for positioning the new technological offerings, and the research presented here can be helpful in such positioning.

In chapter 8, Kaufman and Lane discuss technologies and the use of time. The effects of increased use of technology in households on people's time utilization are difficult to measure. By accelerating and automating processes, technologies compress time. Furthermore, technologies often facilitate the simultaneous conduct of multiple activities or "polychronic" time use. Kaufman and Lane illustrate how use of technologies can overlap in a multisensory fashion, and how technologies figure in the matrix of resources available to a household. To adequately analyze the use of time and technology, the authors advocate the research methodology of contextual inquiry. This approach permits in-depth assessment of actual technology use in private homes and can thus generate insights beyond those generated by conventional survey methods.

Mundorf and Westin (chap. 9) discuss acceptance and adoption of technology as a function of different factors. Their focus is on individual differences, notably the demographic variables of gender, age, culture; and product features such as content, structure, and user interface of communication technologies. They show that there is generally a male bias in the acceptance of information technology. Such bias, however, can be overcome by adjusting content and user interface to account for differences. Females show less familiarity with television-based technologies, whereas no significant gender difference is found for phone-based technologies. College-age users demonstrate greater interest in such technologies than those in older age groups. To test the influence of culture, Mundorf and Westin report a comparison with Germany. Younger age groups are more conservative regarding technology compared to their U.S. counterparts. In addition, Germans are less familiar with some of the technologies. Differences in computer use, however, are mainly a function of age group: College-age Americans and Germans showed greatest familiarity. User interface variables such as music and color impact the acceptance of an information service, as does content. Music increases liking by both genders, color only by males. College-age users show preferences for entertainment rather than information or transactional functions of a service, which supports the notion of entertainment as the "driver" for information technology diffusion in the home.

Fırat (chap. 10) takes a postmodern angle focusing on the issue of literacy. During the enlightened period, discourse centered around the written word. New technologies have made a new kind of discourse possible—one grounded in visual imagery. Unfortunately, communication based on visual discourse, popular among adolescents, has been monopolized by the entertainment industry. Educators still are rooted in the written word. Utilizing postmodern, visual discourse might improve the interaction between educators and their clientele, and increase adolescent motivation to acquire knowledge and problem-solving strategies. Fırat makes the case for a postmodern literacy based on the skills necessary to understand and create such visual discourse.

Policy Perspectives

Although deregulation has diminished the role of government and increased the role of the market in influencing the availability of and demand for information products and services to users at home, this does not imply that public policy is no longer a determinant of demand patterns. Like the rest of the chapters in this book, the four chapters in this section provide perspectives for understanding, evaluating, forecasting, and managing new information technologies and the services based on such technologies. These chapters, however, view the field from a different vantage point—that of public policy.

Historically, in the United States, telecommunication services have been seen as an essential public utility and therefore have been subject to state-level regulation. In today's deregulated, digitalized, and de-tethered world, the relevance and forms of such state-level regulation have come into question. Based on his experience as the chairman of the Rhode Island Public Utilities Commission, Malachowski (chap. 11) explores the challenges state-level regulators face in these complex, dynamic times. They have to balance the twin, often incompatible, goals of protecting vulnerable service users and transforming the state's communications infrastructure for economic development and competitiveness. Malachowski outlines the approach his agency adopted and offers valuable lessons for regulators in general.

Chapter 12, by Dutton, explores the experiences of electronic service delivery in the public sector of the United States. Dutton takes a comprehensive view of the public sector and includes federal, state, and local government agencies as well as some quasigovernment institutions. Such government efforts for electronic service delivery often provide a laboratory for future private services to come. Underwritten by public monies, available to all eligible citizens or residents, and fully open to public scrutiny, government delivered electronic services are in a sense mass-scale trials. Even if the level of technology does not reflect state-of-the-art practice, the massive nature of these social experiments creates generalizable learning for those designing, delivering, assessing, or critiquing private electronic services. Dutton draws many valuable lessons of this nature and challenges some of the myths that have evolved about what works and doesn't work in the field of electronic services.

Although all economically advanced nations are striving to position themselves at the leading edge of the Information Age, public policy orientations and approaches differ across countries. In particular, France was successful in launching videotex services in the 1980s in spite of the general pessimism that followed the failure of videotex trials of the 1970s. Steinfield (chap. 13) reviews the key elements of the French approach and compares and contrasts these to what has been happening in the United States. Steinfield feels that the U.S. public policymakers have not yet been able to come up with a low-cost, nationwide, universally accessible system for information services.

In the final chapter of this section, Thomas explores the long-term social implications of new information technologies. Major changes are occurring in the development of technologies and these are classified as trends, breakthrough, architecture, and metatechnology by the author. The magnitude of the changes vary: Although trends represent incremental extrapolations of a particular technology, architecture and metatechnologies imply major transformations in the ways we conduct business and organize human activity. Social responses to these changes are also not uniform. From denial and anger, individuals may respond to the changes by making deals and finally accepting them. All of these responses, however, are passive argues Thomas and he encourages, instead, a more proactive view that involves transcendence and envisionment. This proactive view, the author hopes, will lead to design of information technologies that enhance the greater good and facilitate world wide as well as local cooperation.

THE ROAD AHEAD

The Information Society

In the economically advanced countries, the information society is no longer a concept for the future—it is here now. From the viewpoint of the information technology user at home, the emerging industry structure he or she faces is less monolithic than the telecom monopolies and giant broadcast networks of the past but more concentrated than some of the recent ways of obtaining information and entertainment such as independent video rental stores and local libraries. It is reasonable to assume that the emerging systems for delivering information will be quasioligopolistic, at least in behavior if not in terms of industry structure. At the same time, the pressure for variety and novelty, both on the supply and the demand side, will lead to greater diversity in information sources. From the user's perspective, a crucial issue is going to be access to varied information sources: Will such access be heavily mediated by the giant "infotainment" conglomerates that are forming in major economies of the world, or will tomorrow's intelligent networks allow the consumer to bypass the quasi-oligopolies and tap directly into myriad infotainment sources?

To a great extent, the answer to this question depends on how well the infotainment providers, technology developers, policymakers, and consumer advocates understand the needs, behaviors, and preferences of users. The more thoroughly these are understood, and the more widely knowledge about users' perspectives is disseminated, the greater the chances that a diverse, competitive, user-oriented structure will emerge to serve the infotainment user of the next century. In other words, user empowerment depends on the ability to understand and widely project user needs—manifest as well as latent. The chapters in this volume have provided a variety of perspectives on understanding the user and in approaching the user from private strategic as well as public policy angles. We hope this volume will serve to facilitate further work in the area of understanding the demand side of new information technologies.

REFERENCES

All things considered. (1994, January 3), *National Public Radio.*
Auletta, K. (1994, April 11). The magic box. *The New Yorker,* pp. 40–45.

Baldwin, T., & Litman, B. (1992, April). *Maximizing profits in a 400 channel world: The fate of commercial broadcasters.* Paper presented at the Broadcast Education Association Convention, Las Vegas, NV.

Blankenhorn, D. (1993, April 26). Sculley sees future in digital. *Electronic Media,* p. 28.

Brosius, H. B., & Mundorf, N. (1990). Eins und eins ist ungleich zwei: Differentielle Aufmerksamkeit, Lebhaftigkeit von Information und Medienwirkung [One plus one unequal two: Differential attention, vividness of information and media effects]. *Publizistik, 35*(4), 398–407.

Byrne, J. A. (1993, December 20). The horizontal corporation. *Business Week,* pp. 76–81.

Carey, J. (1992). Looking back to the future: How communication technologies enter the American household. In J. V. Pavlik & E. V. Dennis (Eds.), *Demystifying media technology* (pp. 32–39). Mountainview, CA: Mayfield.

Carey, J. (1993). *Lessons from failure: New information technologies yesterday, today and tomorrow.* Paper presented at University of Rhode Island Honors Colloquium, Kingston, RI.

Dholakia, N., and Dholakia, R. R. (1995). The changing information business: Towards content and service-based competition. *Columbia Journal of World Business,* pp. 94–104.

Dholakia, N., & Mundorf, N. (1993). *New information technologies: Panacea or peril.* Kingston, RI: University of Rhode Island.

Dholakia, R. R. (1992). Competition between goods and services: Setting the research agenda. In J. N. Sheth (Ed.), *Research in marketing* (pp. 81–114). Greenwich, CT: JAI Press.

Dholakia, R. R. (1993). The plugged-in home: Marketing of information technology to U.S. households. In P. Zoche (Ed.), *Heraus-forderungen fur die Informations-technik* (pp. 86–100). Heidelber: Physica-Verlag.

Dholakia, R. R. (Ed.). (1994a). *Strategic perspective on the marketing of information technologies.* Greenwich, CT: JAI Press.

Dholakia, R. R. (1994b). The marketing challenge: When services compete with products. In R. R. Dholakia (Ed.). *Strategic perspective on the marketing of information technologies* (pp. 55–70). Greenwich, CT: JAI Press.

Dholakia, R. R. (1994c, May). Even PC men won't shop by computer. *American Demographics,* p. 11.

Dholakia, R. R., Dholakia, N., & Pedersen, B. (1994, December). Putting a byte in the gender gap. *American Demographics,* p. 20.

Dholakia, R. R., Pedersen, B., & Hikmet, N. (1995). Married males and shopping: Are they sleeping partners? *International Journal of Retail & Distribution Management,* pp. 27–33.

Dholakia, R. R., & Venkatraman, M. (1993). Marketing services that compete with goods. *Journal of Services Marketing, 7*(2), 16–23.

Dupagne, M., & Agostino, D. E. (1991). High-definition television: A survey of potential adopters in Belgium. *Telematics & Informatics, 8*(1/2), 9–30.

Eckersman, W. (1991). Users need to adopt telecommuting plans. *Network World, 8*(11), 27–28.

Elmer-Dewitt, P. (1993, April 12). Take a trip to the future on the electronic superhighway. *Time,* pp. 50–55.

The entertainment economy. (1994, March 14). *Business Week,* pp. 58–64.

Fabrikant, G. (1993, May 18). Present at the marriage of phone and cable TV. *New York Times,* pp. D1,2.

Fırat, A. F., & Dholakia, N. (1982). Consumption choices at the macro level. *Journal of Macromarketing, 2*(2), 6–15.

The future of TV. (1993, November 8). *Electronic Media,* pp. 23–31.

Garhamner, M., & Gross, P. (1993). Effects of flexible work hours and technologies in the home on leisure (in German). Bamberg: Sozialwissenschaftliche Forschungsstelle der Universitaet.

Gilder, G. (1994, December 5), Telecosm: The bandwidth tidal wave. *Forbes ASAP,* pp. 162–177.

Haugsted, L. (1993, November 15). What kinds of programming will flesh out interactive nets? *Multichannel News,* p. 34.

Hontz, J. (1995, January 2). FCC ruling gets mixed reaction. *Electronic Media, 3,* 50.

Hornik, J., & Schlinger, M. J. (1981), Allocation of time to the mass media. *Journal of Consumer Research, 7*(4), 343-355.

Kaufman, C. F., & Lane, P. M. (1993). *Time and technology: Acquisition and use of household innovation* (Working Paper: RITIM). Kingston: University of Rhode Island.

Kaufman, C. F., Lane, P. M., & Lindquist, J. D. (1991, December). Exploring more than 24 hours a day: A preliminary investigation of polychronic time use. *Journal of Consumer Research, 18*(3), 392-401.

Keller, J. J. (1993, July 28), AT&T's secret multimedia trials offer clues to capturing interactive audiences. *The Wall Street Journal,* pp. B1, B6.

Kiesler, S., & Sproull, L. (1992). Group decision making and communication technology. *Organizational Behavior and Human Decision Making, 52,* 96–123.

Lieberman, D. (1994, October 10). Tests could pave way for more services. *USA Today,* pp. 1B, 2B.

Maddox, K. (1994, May 16). Poll: 59% in the dark on information highway. *Electronic Media,* p. 8.

Malone, T. W., & Rockart, J. F. (1991, September). Computers, networks and the corporation. *Scientific American,* pp. 128–136.

Media Mania (1993, July 12). *Business Week,* pp. 110–119.

Meyer, S., & Schulze, E. (1993). *Projektergebnisse: Technikfolgen fuer Familien* [Project results: Effects of technology for families]. Duesseldorf: VDI-Technologiezentrum.

Mundorf, N., & Dholakia, R. R. (1994). *Video on demand in the United States: A survey of trials and market potential.* Kingston: RITIM, The University of Rhode Island.

Mundorf, N., Dholakia, R. R., Dholakia, N., & Westin, S. (1995). Orientations towards Technology in Germany and the U.S.: Implications for Marketing and Public Policy. *Journal of International Consumer Marketing.*

Mundorf, N., Meyer, S., Schulze, E., & Zoche, P. (1994). Families, information technologies, and the quality of life: German research findings. *Telematics and Informatics,* 137–146.

Mundorf, N., Westin, S., & Dholakia, N. (1993). Effects of hedonic components and user's gender on the acceptance of screen-based information services. *Behaviour and Information Technology, 12*(5), 293-303.

Mundorf, N., & Zoche, P. (1993). Nutzer, private Haushalte und Informationstechnik [Users, private household, and information technology]. In P. Zoche (Ed.), *Heraus-forderungen für die Informations-technik* (pp. 61–69). Heidelberg: Physica.

Orme, P. (1994). *Interactive and broadband strategy in a changing communications environment.* Ridgefield, CT: P. Orme Associates.

A phone-cable vehicle designed for the data superhighway. (1993, October 14). *New York Times,* pp. A1, D10.

Rebello, K., & Eng, P. (1994, May 2). Digital pioneers. *Business Week,* pp. 96–103.

Reese, S. (1990), Information work and workers: Technology attitudes, adoption and media use in Texas. *Information Age, 12*(3), 159–164.

Roderick, J. C., & Jelley, H. M. (1990). Managerial perceptions of telecommuting in two large cities. *Southwest Journal of Business & Economics, 18,* 35–41.

Rogers, E. M. (1986). *Communication technology: The new media in society.* New York: The Free Press.

Saloman, I., & Koppelman, F. (1988). A framework for studying teleshopping versus store shopping. *Transportation Research, 22A,* 247–255.

Schlender, B. (1994, January 16). What Bill Gates really wants. *Fortune, 131*(1), 34–63.

Sell, M. V., & Jacobs, S. M. (1994). Telecommuting and quality of life: A review of the literature and a model for research. *Telematics & Informatics, 11*(2), 81–97.

Saffo, P. (1993, Autumn). The future of travel. *Fortune* (Special Issue), pp. 112–119.

Steinfield, C., Bauer, J. M., & Caby, L. (1994). *Telecommunications in transition.* Thousand Oaks, CA: Sage.

Still on the line. (1985). *American Demographics, 7*(10), 14.

25 breakthroughs that are changing the way we live and work. (1994, May 2). *US News & World Report,* p. 46–60.

Venkatesh, A., & Vitalari, N. (1986). Computing technology for the home: Product strategies for the next generation. *Journal of Product Innovation and Management, 3,* 171–186.

Venkatesh A., & Vitalari, N. (1987). A post-adoption analysis of computing in the home. *Journal of Economic Psychology, 8,* 161–180.

Waldrop, J., & Exter, T. (1991, March). The legacy of the 1980s. *American Demographics, 13*(3), 33–38.

Walker, J. R., & Bellamy, R. V. (1993). *The remote control in the new age of television.* Westport, CT: Praeger.

Weber, J. (1993, October 31). New TV media buyer's dream, a nightmare to ad agencies. *The Providence Sunday Journal,* p. F1,5.

Westin, S., Mundorf, N., & Dholakia, N. (1993). Exploring the use of computer-mediated communication: A simulation approach. *Telematics and Informatics,* pp. 89–102.

Witte, E. (1993). Was erwarten wir von der Telekommunikation? In P. Zoche (Ed.), *Heraus forderungen fur die Informations technik* (pp. 239–246). Heidelberg: Physica.

Wood, W. C. (1986). Consumer Spending on Mass Media: The Principle of Relative Constancy Reconsidered. *Journal of Communication, 36,* 39–51.

A year of turmoil in communications. (1993, October 14). *New York Times,* p. D11.

Young, K. (1992, August). Teleworking: Home alone. *Telecom World,* p. 27.

Zillmann, D. (1988). Mood management: Using entertainment to full advantage. In L. Donohew, H. E. Sypher, & E. T. Higgins (Eds.), *Communication, social affect, and cognition* (pp. 147–171). Hillsdale, NJ: Lawrence Erlbaum Associates.

Zillmann, D. (1993). Cognitive and affective adaptation to advancing communication technology. In P. Zoche (Ed.), *Heraus-forderungen fur die Informations-technik* (pp. 416–428). Heidelberg: Physica.

Zuboff, S. (1988). *In the age of the smart machine: The future of work and power.* New York: Basic Books.

PART I
Managerial Perspectives

CHAPTER 2
Market Opportunities and Pitfalls for New Information Technology in the Home

Mats Samuelsson
General Instrument Corporation

We must take the current when it serves, Or lose our ventures
—Julius Caesar, Act IV, scene iii

Since the mid-1980s, one home information technology after another has presented itself in the market place. Many of them have been miserable failures and it is easy to see the pitfalls that were overlooked when looking back. After all, who does not remember the RCA video-disc, IBM's PC junior, videotex, smart homes, and other hyped up universal solutions to the home of the future. In fact, each Christmas season brings with it at least one major new consumer electronic product that promises to bring more information technology into the home. In 1993 it was sophisticated (and expensive) 32-bit video games, home theater systems, and multimedia computers. The 1992 season brought us interactive compact discs (CDs), recordable mini-discs, and baking machines! Going back we can find example after example of false starts and even disasters. The RCA video-disc effectively finished off RCA as an innovator and their revered "Sarnoff" Laboratory was spun off and has hardly been heard from since. The home computer craze from the early 1980s caused embarrassment for major computer makers and caused several of them to withdraw from consumer electronics. The pitfalls are many and easy to identify. But if we look inside the home in the mid-1990s, and compare it to the mid-1980s, we find dramatic differences in the area of information technologies. The typical family has a VCR, is hooked up to cable television (CATV), and the cordless phone is answered by an answering machine. The children play with video games, watch the Discovery channel, and are surrounded by electronic toys. Teenagers' stereos have CD players. Many families have or are again seriously considering home computers, especially for their children (Armstrong, 1994). Movies are watched on the home entertainment system, complete with surround sound. Family events are captured using camcorders or in some cases photo CDs.

Instead of a fast and dramatic introduction of the new "ultimate" home information technology, there has been a gradual introduction of information technologies into major and minor areas of the home and the life of its occupants. In some areas, the cumulative impact of such changes has been dramatic. A look at the ownership numbers as of 1993 is revealing: 60% of U.S. households have CATV, 91% a VCR, 59% an answering machine, 49% a cordless phone, 43% own CD players, 44% have stereo TVs, and 9% already own projection TVs (Electronic Industries Association, 1994). Even though the diffusion of home computers has been far slower than predicted, a surprising 37% have a home computer. Most of these information technologies did not even exist in 1980.

Although some of this was forecasted in the mid-1980s, much of it has happened in unexpected ways far different from the promises and hype accompanying the announcements of new breakthrough information technologies. It is as if far below these promises and hype, there has been and will continue to be a slow and steady stream of products and services that introduce new and very sophisticated information technologies into the home (Lohr, 1994).

Without knowing the exact outcome, we know that the home will not look the same in the year 2005. In trying to assess both market opportunities and pitfalls for new home information technologies we must therefore include an analysis of what some of the pitfalls have been in the past as well as look at how technologies actually enter the home and what it is that has driven the opportunities and successes of the past.

PITFALLS FOR NEW INFORMATION TECHNOLOGY IN THE HOME

The present will always look different in hindsight and the initial analysis of past events is not always right and many times conflicting. For example, the initial analysis of many disasters and pitfalls in home information technology has been summarized as the technology "being too early." Pundits claim that the PC junior was too early but that today's personal computer invasion into the home proves that the PC junior concept was the right one. Such retrospective analysis carries the smug satisfaction of a 20/20 hindsight but offers little else. A more thorough look at successes and failures points at something different. Technology alone is seldom successful. There is a lot of truth in the statement "technology looking for a market." Unless technology performs something useful, something that a large number of individuals and families will perceive as desirable or valuable over time, it is likely to land in the dustbin. For technology to be successful, it also has to be embedded into easy-to-understand applications that are straightforward. And instead of a revolutionary and fast introduction, many technologies are only gradually adopted into the home over a long time span, sometimes stretching over 20 years. Some examples of technology successes and failures illustrate these characteristics:

1. CATV is useful because it provides many more TV channels than on-air TV and premium channels let us watch current movies (Eastman, 1993). However, it took 20 years to penetrate 60% of U.S. homes. (Success Factor: Gradual adoption)
2. Video games are enjoyed and played by children (Mandel et al., 1994; Success Factors: Technology embedded into easy-to-understand applications)
3. Compact discs are small and easy to handle and we like the high quality and the convenience. (Success Factors: Gradual adoption, technology embedded into easy-to-understand applications)
4. The answering machine enables people to reach us and we can screen calls. (Success Factors: Gradual adoption, valuable function, technology embedded into an easy-to-understand application)
5. The PC junior did not provide any value. Programs were boring and after a while the computer collected dust. (Failure factors: Technology by itself, no useful application)
6. Videotex was too slow (Alber 1985; Westin, Mundorf, & Dholakia, 1993). In front of the television is not a convenient place to shop for airline tickets; instead, call a travel agent. (Failure Factors: Not useful or valuable over time)

So-called enabling technologies (i.e., products that do nothing on their own but supposedly enable you to do other things) are seldom successful. For example, home automation has been a hard sell and a miserable failure. Equally difficult are products that require sophisticated users, complicated set-ups, or some type of special capabilities to work. When do you find the CATV connection right next to a telephone jack? Since when do people have their homes wired up for complex multiroom distribution of television and entertainment systems? Who has a phone jack in the bathroom?

Sales, marketing, timing, price, and accessible distribution channels are of course important determinants of how successful new home information technologies will be (Carey & Elton, this volume). How can you sell videotex? In hindsight, could you have marketed a personal computer without appealing applications? Can the market absorb several digital recordable media at the same time, and just after people bought their first CD? Can you sell a $800 video game for children? Do people buy toys in electronics stores? As we can see, it is easy to look at past product/service successes and failures and draw conclusions from them. It is harder to try to identify future pitfalls that have yet to appear and forecast what will succeed or fail in the future? The fact that it is hard, however, does not mean that it is impossible and a way of getting started is to take a closer look at some of the parameters that have characterized the actual adoption of successful new home information technologies in the past.

TECHNOLOGY DIFFUSION INTO THE HOME

New home information technology seems to be surrounded by promise and hype. To a large extent, this constitutes a marketing discourse designed to impress customers, media, the distribution and reseller network, and industry constituencies like the investment community and even corporate management (Carey & Elton, this volume). This marketing discourse and the actual adoption of information technology seem to operate on two different planes. The former seems to reflect the need for businesses to announce new radical products as a way to be the first to go after an opportunity. The belief in a first mover advantage and universal breakthrough products seems to be strong enough in high-technology industries to assure a steady stream of such announcements and continuous attempts by individual companies to grab market share by announcing something new early. They thrive on being Techno-Optimists providing a clear path to the future.

The real-life diffusion and adoption pattern of information technology often follows a less certain path (Lohr, 1994). Markets tend to be heavily segmented, the growth rates gradual, and simple, useful, easy-to-use applications that provide immediate value dominate the successes. The initial impact of the technology in the home is usually low, although the cumulative effect over a period of 5 to 10 years often produces dramatic differences in the typical home. In some cases, this change can be revolutionary as in the case of the widespread adoption of advanced video technology and home electronics in the 1990s (Kaufman & Lane, this volume). Real-life diffusion and adoption patterns clearly differ substantially from the vision provided by the Techno-Optimists. Table 2.1 shows the contrast between vision and reality for some important market parameters.

The successes have one thing in common. People found an initial use for them in their daily life. If powerful enough, this initial "wedge" application started to crack open the market and continued to drive the penetration. Failures are clearly marked by a lack of such wedge applications or by weak wedge applications that failed to crack open the market. Examples are the PC junior or Videotex that never found any use (lack of wedge application) and the RCA video-disc that could only offer playback of prerecorded

TABLE 2.1
Techno-Optimism Versus Real-Life Diffusion

	Marketing Hype/ Techno-Optimism	*Real-Life Diffusion and Adoption Pattern*
Market size/type	Mass/universal	Segmented
Market growth rate	Rapid	Gradual
Applications	Multiple/sophisticated	Single/simple initial useful application
Usage characteristics	Versatile and highly programmable	Simple, embedded, easy to use
Impact of technology	Revolutionary impact at introduction	Evolutionary, heterogeneous, gradually adding up to a revolution

discs, not home taping for time-shifting of TV viewing and preservation of favorite programs (weak wedge application wedged out by more versatile VCR). Without the strong wedge application, even initially successful information technology will fail to open up a wider market (Rogers, 1986).

The speed at which markets open up and grow is obviously related to the appeal of the wedge application. However, more traditional marketing parameters like pricing, promotion, and distribution channels play a very critical role in consumer markets. Many initial product offerings are clearly priced too high as if price elasticity curves did not exist and as if the old successful Texas Instruments experience-curve-driven pricing approach, used so successful with electronic calculators (the first microprocessor information technology to enter the home), had never been used. At Christmas time, it is the pricing of information technology that decides the gift giving pattern (Armstrong, 1994). Traditional familiar distribution channels are also more effective in speeding up the adoption of new technologies than approaches requiring changes in buying patterns. Finally, people like to buy tangible things, a challenge for information service providers (Dholakia, 1992, 1994).

Looked at from this perspective, past successes seem obvious in hindsight. Isn't it obvious that people will only buy home PCs in large quantities when they have a number of appealing applications for them! The recordable VCR is so much more versatile than the video-disc player! How could we live without answering machines? The wedge applications are easy to identify. Once the wedge application is there, new features and bells and whistles are often used to penetrate the market. Although many people will not use all the features of a new home information technology they seem to need the reassurance that it is feature-laden, that the features are available at their finger tips and that the features are useful in their everyday life.

From this discussion, it is clear that wedge applications and how the accompanying information technologies are marketed are critical for the success of any new home information technology. Although new wedge applications are clearly critical, identifying them is futile without an understanding of which products and services it will be possible to offer in the next 5 to 10 years (i.e., which wedge applications are even conceivable).

FOUNDATIONS FOR NEW INFORMATION TECHNOLOGY OPPORTUNITIES

When we start to look at new information technology opportunities, we have help from experience, particularly since the 1980s. New information technology is, as the name implies, technology intensive. The underlying technology in many ways limits what it is able to do. In assessing the future, we therefore need to look at trends in the underlying technology evolution because it will form the critical foundation of many new market opportunities. A helpful concept is that of a *technology current* (see Fig. 2.1). This technology current can be thought of as the combined sum of the large number of individual technologies that are being (or will be) utilized in new information technologies for the home. This can incorporate anything from chip technology to new media formats or communication interface specifications.

Each individual technology has its own special characteristics and technology curves. Technology curves can be said to exist when there is a relatively predictable relationship between time and price or performance characteristic of a given technology. Examples are continuous price/performance improvements for chip and memory technology (Technology 1 in the figure), life-cycle curves for CD technology (Technology 5) or S curves for communication technology (Technology 3).

Although the technology current looks linear, it tends to flow in spurts as new underlying technologies are combined to create completely new types of products that then become part of the technology current. The CD, for example, is the result of independent but simultaneous improvements and innovation in chip-processing speeds, laser technology, plastics technology,

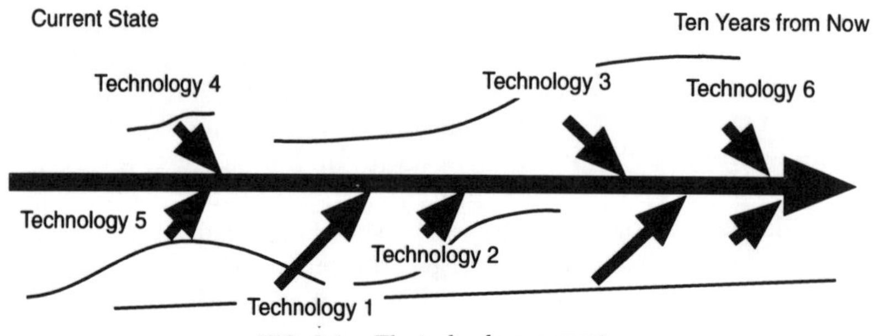

FIG. 2.1. The technology current.

and manufacturing innovations. The CD technology has then been used for a wide variety of applications, the latest being multimedia entertainment programs. Although it is extremely hard to know the exact content and nature of the technology current in the future, many of the technology curves that feed into it can be extrapolated easily. It is also possible to envision some major changes based on the known simultaneous changes in several areas. Some important technology curves on component and system levels and an example of trend shifts that are part of the future technology current are discussed here.

Since the mid-1980s, there has been a 100-fold price/performance increase in computer chip technology. The clock speed is the most dramatic example. In 1980 1 MHz was a challenge. In the mid-1990s, the leading edge is 100 MHz. With 1 million instructions per second you can run slow word processors or spreadsheets. With 100 million per second you can process your home videotapes and use your PC as part of your high quality stereo system.

Computer memory technology is following a similar track. A new little chip type, the digital signal processor, is making inroads into consumer electronics. This has just started in camcorders and PCs. The ability of this chip to process voice, image, and video will most likely revolutionize product functions and human interfaces.

In 1980, cellular handsets cost more than $1,000 and cordless phones cost $500. In the mid-1980s, prices for both are around $50. This clearly has an impact on what future hand-held information technologies can do.

In 1980 the investment cost of installing a 4 kHz voice telephone connection was around $1,000. At that time, this was the cost reference of an important communication delivery mechanism into the home, the telephone. In 1995, it cost the same amount to get a 3 MBit/s digital connection, a connection that can support interactive television and multimedia applications. In effect, the cost per kbit/s dropped from $100 to about 35¢.

The last trend in particular is important for the future because it is influencing the decisions of CATV and telephone companies to upgrade their respective networks so that they can provide digital interactive television and multimedia services. These digital highways into each home will revolutionize what can be offered over them. For the first time, it will be possible to connect the television to the same wire (or fiber) in the wall that the phone is connected to, or the other way around, depending on the perspective taken. It will also be possible to connect video games, stereo systems, and home computers to the same wire.

Right now, the technology current that provides the foundation for new information technologies is going through dramatic changes. Without any doubt, some of these changes will be the foundations for new market opportunities.

MARKET OPPORTUNITIES: DEMAND-SIDE PERSPECTIVE

The difficulty in assessing market opportunities for new information technologies from a true demand (i.e., customer) perspective is that it often turns into a guessing game of what it is customers really want in the future or what the new wedge applications will be that will pry open new markets. These assessments and guesses tend to have one thing in common, they are nearly always wrong (Carey & Elton, this volume). The reason for this is that they tend to rely on simplistic models of how product and service markets really work, confusing the relationship between demand and supply. Although it is easy to show through hindsight that intrinsic customer needs are always met by successful new products, many needs are anything but intrinsic and certainly not predictable (Simson, 1993).

In fact, a case can be made that many needs are created by supply. For example, the availability of multimedia PCs with CD-ROM is clearly a supply concept that was put together because something was suddenly technically possible, a concept that then went out looking for uses and applications and is in the process of finding wedge applications right now. Rather than being demand driven, this is a technology-driven concept powerful enough that it will most likely create a major market (Armstrong, 1994). In 1993, there was no demand or need. In 2005, more than half the population might not want to live without one.

Rather than trying to forecast needs, looking at market opportunities therefore becomes a question of in which information technology areas the interaction of intrinsic needs, need creation, and supply will occur. Which underlying technologies will provide the foundation for new powerful product/service concepts and wedge applications that are strong enough to create new market opportunities?

This puts us close to a supplier perspective, but not a traditional supplier perspective where suppliers through trial and error try to find out what will sell. This perspective is instead based on what it is possible for suppliers to do based on what the underlying current of technologies will support in the future.

MARKET OPPORTUNITIES: SUPPLIER PERSPECTIVE

From a supplier perspective, finding market opportunities for new information technologies in the home then means looking at what it is possible to do and which demands for products and services that can or will materialize in the next 10 years based on this. In doing this, some things are clear. New products will continue to appear on the market and new suppliers will try to create new markets. Customers will continue to buy information technology, picking and choosing from what is offered. The truths of economics driving supply and demand will continue to be valid, despite the fact that they are often forgotten. Some examples of this are as follows:

1. Supply Drivers. Every company that is currently selling any type of information technology in the residential market will want to continue to do so 10 years from now! Some that are selling information technology today will not be around in 10 years. In 10 years there will be companies around that do not exist today. The same three statements can be made about products.
2. Demand Drivers. As the trend has been over the last 25 years, each home unit will spend a little more, on average, each year on home information technology. Technology evolution means they will get more for their money. Price/demand elasticity curves are applicable in this market like in any other.

Apart from these drivers, wedge applications clearly play a key part in bringing new information technology into the home. It is, after all, the appeal of these that will initially give the product or service a chance. As stated earlier, we know pretty much what the technology current will bring us in the next 5 to 10 years. We also know the present state of information technology in the home. What we don't know is what the state will be 10 years from now. Nor do we know which wedge applications that will enable which technologies to take hold. However, we do know that a continuous stream of potential wedge applications will be developed by the innovative use and combination of the foundation technologies discussed earlier. An interesting analysis of innovation and the interaction between the external technological environment (the technology current) and the external market environment is provided by Marguis (1982). That is, after all, the result of the supply drivers. We also know that customers will buy information technologies and that both use and price are important. These are the demand drivers.

Knowing which technologies will actually make it, is anyone's guess. The telecom and CATV industries hope that one of the applications will be interactive television and multimedia services (Fabrikant, 1993). The home electronics industry hopes that it will be new advanced entertainment systems or high definition television. The appliance industry hopes that it will be new smart appliances. The nation's builders hope that it will be home automation systems. The video game industry hopes that it will be new advanced video games. The computer and software industries hope that the home computer will play a key role and that operating systems will be prevalent in video set top technologies. Needless to say, some of these will come true, others not. The successes will clearly depend on how powerful the wedge applications will be. In some cases the penetration will be high, in other cases few homes will opt for the technology. Figure 2.2 illustrates the flow of the technology current and the wedge applications that are generated over time.

Wedge applications are typically the result of spurs of innovation that create excitement among users or based on the fact that a clear market need can suddenly be met because of progress in the technology current. Video games are examples of the former, whereas the VCR and the CD player represents an example of the latter type. Wedge applications can also be origins of obsolescence for current products (CD vs. record), new ways of doing things (home entertainment systems) or something completely new (car navigational systems). Which wedge applications (and the technology associated with them) succeeds is dependent on many things. In some cases, the application itself is powerful enough to succeed. This is clearly the case with answering machines. In other cases success is dependent on marketing, standards, or complex market coalitions as in the complex VCR business where the emergence of the video rental store played a critical part. Quantitative changes in the technology curve can also result in qualitative improvements. This concept will be tested to its limit as new interactive television and multimedia services try to become more successful than their low-speed predecessor, videotex. So what new information technology will make it into the homes during the next 10 years?

Let us look at the underlying technology current. Although there will be one or two types of completely new technologies utilized in products and services 10 years from now, the majority used will be extensions and

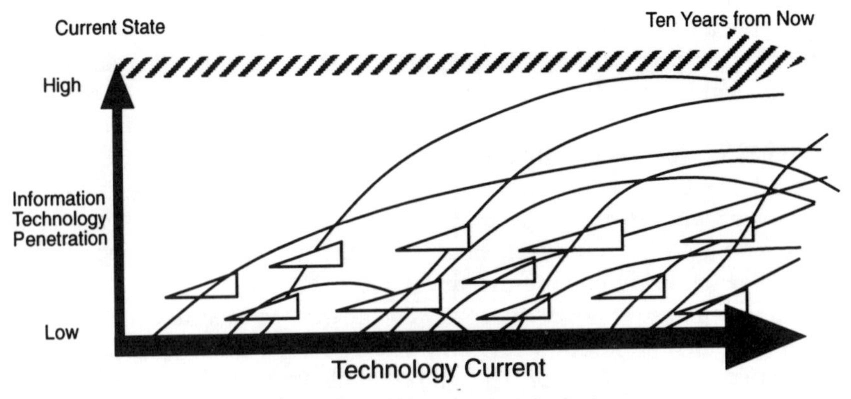

FIG. 2.2. Information technology in the home.

improvements of technologies that we have today. The difference lies in that the capabilities offered by each technology will improve dramatically. This means that completely new uses will be possible and the resulting products and services will appear through invention and innovation. There is room for completely new wedge applications ("25 Breakthroughs," 1994).

The major upgrades in CATV and telco networks under way will also open a tremendous door for innovation in completely new home electronics/ communication areas. As the innovation cycle using these capabilities starts, market opportunities and demand for new information technologies will come right with it. Potential newcomers in the technology current will add to this:

- Intuitive user interfaces using video/graphics/sound.
- Dramatic increase in communication services and programming coming from the combined CATV/telco outlet. Customized video, audio, game and data services.
- New types of hand-held devices based on increased miniaturization and communication capacity (making possible anything from miniature portable video phones to complete hand-held stereo and video systems).
- One or two new "formats" for video, data, or sound programming. A silicon information cube that replaces today's tape or CD storage media with memory. Such a unit would have no moving parts.

CONCLUSION

As for thinking about what the outcome will look like (i.e., which wedge applications will succeed) a good analogy is to think back 10 years about one of the first video games, "Pong." The game consisted of a simple ball bouncing back and forth on a black-and-white screen. Using a simple lever that moved a white square up and down on the side, the player was supposed to intercept the ball and bounce it back. At the time, "Pong" was the state of the art, certainly a far cry from today's realistic video games. A whole video game industry has emerged since the 1980s, satisfying the demand for new and creative games, an industry that is now exploring "virtual reality." In assessing the future, we can think of today as "Pong." Only then can we even envision the opportunities that new information technologies will bring 10 years from now. As to who will actually capitalize on these? The only thing we know is that the opportunities are there. The technology current continues to serve a steady stream of predictable and new underlying technologies. The ventures, business opportunities, and wedge applications, coming out of this current remain to be defined. For the suppliers of new information technology, realizing the new opportunities means "taking the current when it serves," otherwise the "ventures will be lost." The result will, no doubt, be some unexpected and exciting information technologies in the home 10 years from now.

REFERENCES

Alber, A. F. (1985). *Videotex/Teletext.* New York: McGraw-Hill.

Armstrong, L. (1994, November 28). Home computers. *Business Week,* pp. 89–94.

Dholakia, R. R. (1992). Competition between goods and services: Setting the research agenda. In J. N. Sheth (Ed.), *Research in marketing* (Vol. 11, pp. 81–114). Greenwich, CT: JAI Press.

Dholakia, R. R. (1994). The marketing challenge: When services compete with goods. In R. R. Dholakia (Ed.), *Strategic perspective on the marketing of information technologies. Advances in telecommunications management* (Vol. 4, pp. 55–70). Greenwich, CT: JAI Press.

Eastman, S. (1993). *Cable/broadcast programming.* Belmont, CA: Wadsworth.

Electronic Industries Association. (1994, January). *Consumer electronic U.S. Sales.* Washington, DC: Marketing Services Department, Electronic Industries Association.

Fabrikant, G. (1993, May 18). Present at the marriage of phone and cable TV. *New York Times,* pp. D1, 2.

Lohr, S. (1994, January 3). The road from technology to marketplace. *New York Times,* p. C11.

Mandel, M. J., Landler, M., Grover, R., DeGeorge, G., Weber, J., & Rebello, K. (1994, March 14). The entertainment economy: America's growth engines: Theme parks, casinos, sports, interactive TV. *Business Week,* pp. 58–64.

Marguis, D. G. (1982). The anatomy of successful innovations. In M. L. Tushman & W. L. Moore (Eds.), *Readings in the management of innovation* (pp. 42–50). Boston: Pitman.

Rogers, E. M. (1986). *Communication technology.* New York: The Free Press.

Simson, E. V. (1993, Fall). Customers will be innovators. *Fortune* (Special Issue), pp. 105–106.

25 Breakthroughs that are changing the way we live and work. (1994, May 2). *U.S. News & World Report,* pp. 46–60.

Westin, S., Mundorf, N., & Dholakia, N. (1993). Exploring the use of computer-mediated communication: A Simulation approach. *Telematics and Informatics, 10*(2), 89–102.

CHAPTER 3
Forecasting Demand for New Consumer Services: Challenges and Alternatives

John Carey
Greystone Communications, Dobbs Ferry, New York

Martin Elton
New York University

> Cable TV is about to become America's National Highway of communications."
>
> —Ralph Lee Smith (*The Wired Nation,* 1970)

Forecasting the demand for new consumer communication and information services has always been problematic. The past century is littered with erroneous forecasts and predictions. Some forecasts have seriously underestimated the demand for services; most have overestimated demand (Klopfenstein, 1989). Examples of underestimating demand have included the telephone, VCRs, answering machines, and cellular telephones. In the 1870s, the telephone was dismissed initially by Western Union as a toy that would have little demand (Brooks, 1975). A century later, AT&T foresaw little interest in wireless telephone calls, although it subsequently acquired one of the largest cellular telephone companies after it became apparent that there was demand for the product. VCRs and answering machines have also outpaced early forecasts of demand (Zangwill, 1993).

It has been more common for forecasts to overestimate the demand for new services. Table 3.1 shows a range in forecasts by six different groups for househould penetration of videotex. The forecasts were made in the early and mid-1980s and predicted penetration levels for videotex in 1990. The average forecast predicted a penetration level of 11 million households. By 1990, approximately 1 million households subscribed to a videotex service. In other words, the error in the forecasts ranged between a factor of 7 and a factor of 20 or more.

In some cases, wildly optimistic forecasts of demand have proven to be embarassing—actual penetration of the technology was zero. For example, AT&T forecast that 1 million picture telephones would be in use by 1980, and 2 million by 1985 (Noll, 1992). However, picture telephones failed in

TABLE 3.1
Market Projections for Penetration of Videotex in U.S. Households by 1990

Group	*Projected Penetration of Videotex (Millions of Households)*
Advertising Age	6.6
AT&T	8
International Resource Development	9.8
IFTF	11
Strategic Inc.	4–12
Southam	20–25

Source: Thomson and Bowie (1986).

the marketplace and achieved virtually no penetration. Similarly, Link Resources (cited in *Cablevision,* December 19, 1983, p. 30) forecast in 1983 that 1.8 million U.S. households would subscribe to direct broadcast satellite services (DBS) by 1985. However, DBS proved too costly to launch and there were no subscribers to the service in 1985.

The poor track record in forecasting demand for new consumer communication and information services has not fostered a cautious approach by those who do forecasting. Each day, major newspapers and trade magazines publish new upbeat forecasts, especially for technologies associated with the information superhighway. A study commissioned by Northern Telecom and Bell South has forecast that video dial tone will generate $30 billion in revenue from new eletronic information services by the year 2000 (Northern Telecom, 1992). Another forecast provides precise penetration levels for video-on-demand by lifestyle and demographic characteristics of consumers (Lohr, 1994). And before a new generation of DBS was even launched, groups were forecasting year-by-year penetration levels of millions of subscribers (MacDonald, 1993).

How do they know? In each case just cited (video dial tone, video-on-demand, and DBS), the forecasts were made before a single customer had purchased a single product. Yet, the forecasts were made by or for major corporations that are using the information to invest billions of dollars. Presumably, they have sound reasons to rely on the forecasts.

In mature, well-understood consumer markets, a small minority of product launches are successful, around 1 in 10. A few successes compensate for many failures. In the field of information and communication technologies, we should expect a similar phenomenon. It is likely to make business sense at a corporate level and economic sense at a national level if new products and services are launched into the marketplace even if the odds are against them. Clearly then, there is a need for forecasting methods that do not implicitly assume that such products and services are bound to establish themselves in the foreseeable future.

We explore several issues associated with forecasting the demand for new informations technologies and services, including who does forecasting, why do companies and government policymakers sponsor forecasting studies, what techniques do they use, how reliable are these forecasts from both a theoretical and a practical perspective, is there a better way to forecast consumer demand, and what are some alternatives to forecasting as a planning tool?

It is important to note that we are emphasizing forecasts of new information services and technologies. We define *new* to mean that the service has not yet established itself in the marketplace. In this sense, residential picture telephony has been a new service for three decades, whereas cellular telephony, which has been around only half as long, is no longer new. There are many techniques—some more reliable than others—for forecasting demand and revenue over the next few years from products that are well established in the marketplace. New information services and products pose a special problem because they have no track record.

FORECASTING MOTIVES AND PLAYERS

Forecasting is often undertaken to help secure financial and organizational support for a new product or service. The support may be needed internally (e.g., from a corporate board) or externally (e.g., from the investment community). A group that sponsors forecasting activities may look to the forecast to help make a decision about whether to proceed with a new service and, in this sense, is neutral about the outcome of the forecast. Often, however, the decision to proceed with the product or service has already been made by the group sponsoring the forecast. Under these conditions, the sponsoring organization is likely to want a favorable forecast and it is motivated to find a group that will deliver the results it wants. Such groups are not hard to find. Alternately, a sponsoring organization may suppress results from a study that yields a conservative forecast. We are aware of a number of cases in which this occurred.

In the case of government agencies, forecasting studies are often commissioned to assist in long-term planning and to inform the development of policies and regulations for industrial groups such as telephone and cable companies. In the government environment, there also appears to be an inclination to support studies that will yield optimistic forecasts: A member of Congress or a state legislator may have taken a position favoring the deployment of certain technologies, committee staff may see future job prospects in the industry and, equally important, optimistic forecasts imply increased revenues from taxes on the industries being studied. An additional problem in some government studies is a tendency to request forecasting data that no legitimate forecasting organization could possibly provide. For example, a request-for-proposals (RFP) from one federal agency in the late 1980s asked for a forecast of penetration levels for all present and future information/entertainment services by race, income level, and other demographic characteristics of households—25 years into the future.

Forecasting is done by a broad range of commercial groups, nonprofit organizations, and some in the academic community. Many are qualified professionals who employ scientific methodologies to do forecasts (the challenges for qualified professionals and the limits of legitimate scientific methodologies are discussed subsequently). However, others are not trained researchers and they often employ highly questionable methodologies in producing their forecasts. The issue of training and qualifications for forecasters as well as the legitimacy of many methodologies that are used have received little attention. One reason for this is the proprietary nature of many forecasting studies. The studies are commissioned by a private organization and the full report is considered proprietary to that organization. Also relevant here are studies which while not commissioned by a private organization, are sold to them on a multiclient basis for a high price. These studies (i.e., proprietary studies of a single organization and multiclient studies) are never released to the press or the academic community. Instead, a press release is prepared with major findings from the study. However, the study itself and the methodology employed are not subject to the scientific scrutiny that a university or government study would receive. As a result, weak or even fraudulent research can pass under the guise of a press release and be published in major newspapers and trade magazines.

FUNDAMENTAL CHALLENGES IN FORECASTING DEMAND FOR NEW SERVICES

Assuming the best of intentions by a sponsoring organization and a rigorous scientific approach by qualified researchers, the challenges associated with forecasting demand for new services are significant.

Complexity of the Diffusion Process. One of the first problems encountered in trying to forecast demand is the complex environment in which new services are launched. The diffusion of a technology or service is based not only on the desire or demand for the product by consumers but on public policies that may affect whether or how the product is offered, industrial policies such as standards, marketing, pricing and competition (Dholakia, Bakke, & Dholakia, 1991). How can a forecasting methodology account for all of these factors and the complex interactions among them? In addition, any or all of these elements can change year to year, adding to the challenge of a multiyear forecast. No one can know what factors will be at work in the marketplace 3, 4, or 5 years after a product or service is introduced. Further, Arthur (1990) has demonstrated that serendipity and luck are part of the diffusion process. For example, a new service may receive very favorable publicity from a major magazine or TV program, which in turn generates additional publicity and brings public attention to the new service. Or, a company may strike a deal with a major distributor that brings the product before a wide public. This process is characterized as positive feedback by Arthur—certain serendipitous advantages early in the diffusion of a product lead to other advantages in a positive feedback cycle that can help to establish a product or service in the marketplace. Arthur argued that positive feedback is a common phenomenon in the diffusion of new information and entertainment services.

After recognizing the complexity of the diffusion process, we can turn to some common forecasting techniques and models. Each has its own set of limitations.

The Seductive S Curve. The starting point for some forecasts is the fact that, through time, demand follows the S-shaped curve characteristic of epidemiological studies. It starts slow, at some point develops rapid momentum, and eventually flattens out as saturation level is reached. The essence of the approach is to estimate the saturation level, to estimate when it will be reached, and then to fit an S-shaped curve to it, perhaps borrowing a particular curve from some past product or service that shares some attributes with the entity being forecast.

This avoids two ways to fail: having an unrealistically shaped curve for the progression of demand through time (which is an unlikely mistake whatever method is used) and having an unrealistically high level for demand at saturation. The trouble is that other ways in which it is even easier to fail are left open. In particular, the service may never come near to its projected saturation level; there may be little basis for estimating when, if ever, it will get to the point where the curve starts rising sharply; and there may be little basis for estimating how long it will take to get to whatever maximum it does achieve.

The One-Eyed Man Is King. The Delphi method and its variants have proved popular as means of forecasting demand for new products and services. The primary aim of the technique is to derive a consensus forecast from a group of experts (Linstone & Turoff, 1975). Secondary attributes include allowing the experts to educate one another further in the process and weighting initial inputs to reflect varying degrees of expertise. So the technique is not in essence a forecasting technique at all; it is a technique for group problem solving (Martino, 1985).

The Delphi technique has some value. Interaction among the experts in the Modified Delphi Technique can certainly produce insights that are of value to marketers. The different versions of the technique bring some discipline to guesswork. And the technique has a certain credibility: How can one do better than to rely on acknowledged experts in the field?

The trouble is that the credibility is undeserved. There is no evidence that experts' guesses are better than anyone else's. Indeed, there is some reason to expect that they may be worse, because they are often biased by personal interest in the success of the technology or service. Besides, if expertise is of value here, why is it necessary to develop group processes for reconciling enormous initial differences among the experts? One can also ask how a person can be an expert in something that is so new? Where should one have turned in the 1960s to look for experts for a Delphi study about residential picture telephony? Or in the 1970s, if one had been concerned about videotex? Expertise in the technology is surely irrelevant. Experts in human behavior? What kind of behavior and from what disciplinary perspective? There is an enormous selection to choose from.

Gilt by Association. Another widespread and seductive forecasting method involves deriving a forecast for the total demand in an established market that will contain the new service and then multiplying this large number by an estimate of the market share of the new service. So demand for videotex was seen as a percentage of consumers' expenditures on information gathering (note the misconception of how videotex would provide value to its users); demand for teleconferencing was seen as a percentage of the projected number of in-person and electronic business meetings; video-on-demand is being considered as a percentage of expenditures on cable television and videotape rentals/purchases. Forecasting the base market is not the problem. There is a reasonable basis of understanding and statistics so one is unlikely to go astray. The problem lies in obtaining a static estimate, let alone a dynamic forecast, for the market share of the newcomer.

Some forecasters appear to avoid the problem by offering a seemingly conservative forecast. They estimate a small fraction for market share, which appears at first sight to be highly cautious. But a small fraction applied to a large base can yield a number that looks quite respectable. For example, if only 10% of revenue from cable television and videocassettes rentals/sales in 1996 moved over to DBS services, it would yield $4 billion in revenue. This is quite impressive. However, the possibility that the fraction might be zero, thus yielding a forecast of zero demand for the newcomer, must also be taken into account.

Other analysts try to adopt a scientific approach to estimating the market share. A classic example was the European telecommunications administrations' (CEPT) attempt in the 1970s to forecast future demand for teleconferencing in business and government (Tyler, 1978). They disaggregated business meetings into different classes and used the results of controlled laboratory experiments on media effectiveness to derive for each class an estimate of the fraction of meetings for which users would in the future substitute different forms of teleconferencing. So the fractions were derived from models of user choice, in this case normative models of the rational economic man or woman, informed by the results of an impressive program of psychological research. However, the forecasts were duds because it was too early to develop and test descriptive models of the choice process—a point to which we return in the following subsection.

It should be noted that this approach to demand forecasting would almost certainly have yielded similar, positive forecasts for 3-D movies in the 1960s, residential picture telephony in the 1970s, and videotex in the 1980s.

Bottoms Up. Rather than work downward from a larger market to the one of interest, other analysts attempt to proceed in the other direction. They start with the individual using or purchasing unit (an individual or a family) and, with widely different degrees of sophistication, they build a model of its choice behavior in a market in which the new service is present (Ben-Akiva, 1990). For different values of its parameters, the model represents the behavior of different kinds of individuals or families that will be present in the marketplace. The model may be expressed mathematically or it may take the form of a computer simulation. Either way, aggregating the individual decisions to obtain estimates for the market as a whole is methodologically straightforward, although extensive survey research may be needed to obtain the necessary estimates of how many purchasing/using units there are of each kind. Further, there may be considerable uncertainty about estimates of the values of certain variables (e.g., the future price of competing products). Here too, the problem lies in the model of choice behavior.

It should be noted that such a model cannot be properly tested if one is dealing with a new service. Nevertheless, using historical data, one can, and certainly should, test the generic model to see how well it would have forecast demand for comparable new products and services in the past. It is, of course, important that the latter should include failures as well as successes. Notice, too, that such models are likely to require forecasts of other variables such as disposable income, interest rates or price levels. Our ability to accurately forecast these economic measures is well known and does not lend confidence to this technique.

Clearly, the construction of a valid model of this kind requires us to understand, at an appropriate level, how the user will derive value from the product or service in question. One would not have developed an adequate model for the telephone at the time when it was perceived as a device for transmitting concerts and church sermons (these were among the applications tested in the 1880s). Nor for the personal computer when it was seen as a tool for processing numbers rather than words. Nor for videotex when it was seen primarily as a service for providing information from data banks and its significance for the exchange of information among users was not properly appreciated.

When new telecommunications products or services are launched into the residential or business market, the question is often whether they can reach take-off speed before they run out of runway. For those that succeed, the applications that turn out to provide sufficient value (take-off speed) are often quite different from those originally envisaged by marketers and their supporting casts of consultants and experts from academe. And by the time these applications are found, demand forecasts may no longer be necessary.

Learn From the Pioneers. The leading edge forecasting technique is used primarily in business settings but it can be adapted to residential products and services. This technique develops forecasts based on the experience of pioneering early users of new systems. Unfortunately, in many cases this level of experience may not exist for new services. There are exceptions—occasions in which useful experiences with a new product or service are available to research before a service has become established. For example, AT&T's trial of Picturephone® within the criminal justice system in Phoenix, Arizona during the 1970s was a success from the users' perspective (Elton & Carey, 1980). However, the company concluded (presumably) that it could not offer the service at an affordable price for a small niche market. Nevertheless, the experience suggested that this might well be a viable niche for some future less costly service. Indeed, this application did emerge successfully in the decade that followed. Such insights can be valuable in making qualitative predictions, although it is not clear how they might be used in making quantitative forecasts.

One of the challenges for those involved with forecasting is the identification of such exceptions (or, in some cases, the creation of exceptions through field trials and demonstration projects) and interpreting their meaning. For new residential products and services certain types of users may provide a useful early experience base (e.g., male teenagers, consumers of adult entertainment, and other early adopter groups). At the same time, care must be exercised in applying what is learned from these groups to other segments of the population.

Boundary Markers. There is a family of techniques, of which the historical analogy method and income and expenditure analysis are key members, which can be useful once a service has started to establish itself. Some of them can be used with care at an early stage. The care is needed because they assume the market success of the product or service in question. Consequently, all they can do for a product or service that has not reached this point is to provide an upper bound to a forecast.

The historical analogy method is based on striking regularities in the time-series data on early sales growth of past communications products or services (Hough, 1979, 1980). It provides a means of making projections using only the first few years' sales data. However, these data are not generally available for new services as we use the term.

Income and expenditure analysis is based on constancies or trends in the proportion of a household budget that is spent on certain forms of consumption (e.g., entertainment and information; McCombs & Eyal, 1980). It is, therefore, a more disciplined variant of the method described earlier in which one estimates demand for a new product or service as a fraction of the total demand in a larger market. Unless used to provide an upper bound, it is subject to the same underlying problem of estimating what if any value, other than zero, the fraction should take. In this sense, it is not really a forecasting technique.

Other regularities, too, can be employed to obtain upper bounds. For example, spending patterns for information and entertainment services can be translated into a percentage of weekly household income. This permits comparisons over time for products and services that were offered at very different prices. It also allows an analyst to track prices in terms of household income at various penetration levels for a technology or service (Carey, 1993). Although useful in understanding historical trends and helping to rule out unrealistically high forecasts, this form of analysis cannot help in estimating whether or when a new product or service will take off.

Historical analysis has another role—as an alternative to forecasting in media planning. This role is discussed later.

Focus on Time. A forecast of demand provides an estimate of demand (e.g., number of units sold or sales revenues) for any point within some future interval of time. This focuses attention on the domain of demand. It would be useful, however, to focus initial attention on the domain of time. For a new service, the key question is how long it will be before it is *established* in the marketplace. (By established, we mean that there is a very low probability that it will be withdrawn for lack of demand.) The possibility that a product will be withdrawn is often seriously underestimated, as is illustrated by picture telephony, video conferencing, videodiscs, and videotex. An earlier example is provided by facsimile transmission (fax), which took many decades before it was established in the marketplace. Operating over telegraph wires, it was introduced before the telephone. Facsimile transmission was withdrawn in the 19th century, re-introduced in the 1930s, and withdrawn again (although two niche forms of facsimile service were established in this era—news photos by wire and meteorological maps to/from ships), re-introduced in the 1950s and withdrawn, and finally became established in the 1980s. Sometimes, however, a new service surprises the pundits and takes off earlier than had been expected: Cellular telephony and answering machines in the U.S. provide examples.

Rather than a conventional forecast, a more useful estimate for a new service would be a plot against time of the probability that its demand will have reached some threshold level. This would draw specific attention to the fact that, at any point in the foreseeable future, the demand might be insignificantly low.

Sometimes the threshold to adopt would be determined by the decision that needed to be taken. Often, decision makers have to choose between only a small number of options (e.g., whether or not to launch a new service in the near future). Analysis may show that, if demand would be above some threshold level by a certain time, the better decision would be to launch, otherwise not. The key question then becomes how probable it is that this demand level will be attained.

Dark Horses and Mugger Technologies. When considering the revenues from new applications and services that contribute to return on investment in significant upgrades of a company's infrastructure, the forecasting problem acquires two other dimensions. One is the risk of entirely overlooking new services that will succeed in the not too distant future. This is not a matter of developing too low a forecast for a service, but one of failing to realize that there is a viable service or application that needs forecasting in the first place. An example is provided by failure to predict the success of a variety of new touchtone services when rotary dialing was to be replaced. When an upgrade is justified primarily by greater efficiency in the provision of established services such oversights may not matter. In the context of fiber-to-the-home or fiber-to-the-curb, they probably matter a great deal.

When considering infrastructure, the other very difficult problem is that even though a new service may establish itself, some other new technology may siphon demand away from the enhanced capability in the infrastructure. Bandwidth compression technology has made it possible to offer some of the services touted by the early supporters of Integrated Broadband Networks (IBNs) at rates of 128 to 384 kbit/s (i.e., within the compass of basic rate integrated services digital network [ISDN]). This should have been an easy mistake to avoid. For an upgraded infrastructure, it becomes necessary to forecast not only the demand for a new service, but the proportion of the demand that would be carried on the infrastructure in question.

A General Formulation. The future demand generated by a new service can be expressed as:

$$D_i(m) = \sum_{n=1}^{\infty} p_i(n)d_i(m/n)$$

where $D_i(m)$ is the demand for new service i in Year m; $p_i(n)$ is the probability that new service i, "takes off" in year n; and $d_i(m/n)$ is the demand for new service i in year m, given that the service takes off in year n.

We prefer this model as a basis for forecasting because it makes explicit the key uncertainty: When will the demand "take off?" In assessing the trust one should place in a forecast, probably the most important consideration is whether the estimate of the time until this occurs is realistic. Experience shows that in the past experts have sometimes erred by a matter of decades.

Research on past time-series data for both successful and unsuccessful new services would help in defining what should be meant by "taking off." One's objective would be to define this as attaining the lowest level of sales and penetration at which one could have confidence that the service would not be withdrawn because of subsequent lack of demand. If it were set too low, growth might dwindle and serious errors of overestimation would result. If, however, it were set too high, there would not be comparable errors of underestimation. This implies that, for a robust forecasting model, there should be a statistical bias toward setting the take-off level on the high side.

When considering the total demand for all new services made possible by an upgrade of the infrastructure, one would be concerned with

$$D(m) = \sum_{i \epsilon S} \sum_{n=1}^{\infty} p_i(n)d_i(m/n)$$

where $D(m)$ is the total demand for all new services in Year m, and S is the domain of all new services made possible by the upgrade.

As noted earlier, another problem arises: overlooking elements in the Set S.

PRACTICAL PROBLEMS IN FORECASTING DEMAND

There are a number of practical problems associated with conducting studies that produce demand forecasts. These include a series of language issues and how researchers provide an experience of the new technology or service to individuals who will provide the feedback that forms the basis of a forecast.

Language and Definitions. When discussing forecasts of the demand for new services, imprecision in the use of language is more of a problem than one might expect. A classic example was the forecast offered by a leading expert at the start of the 1970s. He predicted that the cost of long distance telephony would have fallen by a factor of 10 by the end of the decade. He was referring to the cost of the trunk portion of the transmission. It did not take long, however, for the prediction to be applied to the price of end-to-end service. Because the latter had to cover the costs of the local portion of the transmission and the costs of switching, neither of which was falling at such a rapid rate, the prediction was transformed into nonsense. However, this error in precise use of language did not affect the popularity of the forecast for a few years.

A second problem is the disappearance over time of some statements associated with a forecast (e.g., qualifying assumptions). Careful researchers who attempt to come up with estimates of the future values of market variables normally attach to them explicit statements of the key assumptions used in their derivation. However, it rarely takes much time for exciting forecasts to take on a life of their own no longer fettered by boring assumptions.

Other problems of language are less obvious. For example, there is a common problem with slippery definitions in forecasting studies. When videotex first appeared in North America, its supporters were quick to differentiate the new service from existing online services such as Compuserve. Videotex is different, we were told: Its page format is much more user friendly; it supports color and graphics; and it is organized for tree and branch searching. A great deal of money was spent promoting the new concept and its name became a valuable property. Not for the first time, however, the newcomer failed to live up to the hype. It then became convenient for its supporters to broaden the definition so that preexisting services could be encompassed and a more respectable penetration claimed. Not surprisingly, those associated with earlier services did not complain. Why should they object to the sex appeal conferred by the new term? In similar fashion, the term *video conferencing* grew to encompass the combination of audio conferencing and freeze-frame television. And it seems that a similar fate is befalling ISDN.

Another problem concerns the distinction between a specific product or service and the class of all such products or services: for example, the distinction between AT&T's Picturephone® and the class of all picture telephones. Clearly the probability that, in the marketplace, the class as a whole will succeed—which requires only that at least one of its members should succeed—is very much greater than the probability that any single member of the class will succeed. Ignoring this distinction has provided another cause for confusion.

The central issue here is the clarity of definitions about what is included in a forecast and what assumptions are built into the forecast. If there is a lack of clarity or if assumptions fall by the wayside over time, a reader can be misled about the relevance of the forecast and the forecasters can engage in revisionist history, claiming that the orignial forecast was a success by subsequently re-defining what services or products were included in the forecast.

Service Experience and Validity of Demand Indications. Many forecasts are based on indications by a sample of the public that they would buy a product or use a service. The critical issues are whether they have sufficient experience with or knowledge of the product to make an informed response and whether the verbal response is a reliable indicator of future purchasing behavior.

In the case of products that are well known to the public, the issue of how to pose the question is straightforward (e.g., "Do you plan to buy a new car in the next 6 months?"). Although respondents may overestimate or underestimate the likelihood of buying a car in the next 6 months, at least they understand the question and can provide a reasonably informed response. The issue is much more complex when people have no experience with a product and, indeed, may not have any understanding about what the product is or what it can do. This is the case with many new information technologies and services (e.g., interactive television).

In response to this problem, researchers have employed a number of techniques. One of the most common is to use a standard telephone survey and try to describe the new product or service over the telephone. This technique is the least expensive way to approach the problem and probably the weakest. It tends to inflate positive responses since a verbal description can emphasize positive attributes and respondents have no experience with the product that would allow them to understand potential negative attributes.

A second technique is to intercept individuals in a mall setting and bring them into a room or recruit a large group into a theater setting and show them still photographs, drawings, or a simulation of the service. This is done when the product itself is not yet ready to be used or if it would be very expensive to bring the technology into the research setting. The same problem arises here: Respondents don't really experience the product or service and verbal descriptions that accompany the drawings or simulation tend to emphasize positive attributes. With little chance to understand potential negative attributes, respondents tend to report high interest. A major, multiclient propriety study in the early 1980s used a simulated videotex home banking and shopping service in a theater setting. The results were extremely positive as were the forecasts derived from the study. They indicated that a majority of consumers would readily subscribe to electronic home banking and shopping services. The results were used by a number of large corporations to develop and launch these services. However, when the services were introduced there was very little interest by the general public.

A somewhat better technique is to allow a sample of the public to try the service or product in a laboratory or field setting. Here, people can experience the product for 20 or 30 minutes or longer and provide a more informed response about their potential interest. Although this is clearly better than responses based on a verbal description, it is subject to a novelty effect that often lasts for several weeks. That is, when people first start to use a service, they often respond positively based on the "novelty" of the product and use it more than they will subsequently (i.e., after several weeks of experience with the service).

One way to overcome the novelty effect is through a field trial (i.e., placing the product or service in dozens or hundreds of homes and tracking usage over time as well as indications from users about their willingness to pay for the service after the novelty effect has worn off). There is an added benefit of generating actual experience with a service that can be used for forecasting purposes.

Field Trials. Field trials and market trials have a number of benefits but they are not without their own set of limitations. First, they are generally expensive. The major videotex trials in the 1980s cost tens of millions of dollars. The interactive television trials in the period 1994 to 1996 are certain to cost more than the videotex trials of the last decade. Second, the research findings they are intended to provide can be affected negatively if equipment does not work properly, as happens often when the technology is new. Third, they require a great deal of preparation and a significant commitment of personnel. Implementation requirements for the field trial are often underestimated. Further, they must run for a fairly long period of time to overcome novelty effects and generate sufficient usage/purchasing behavior for demand forecasting purposes (Elton & Carey, 1980).

In addition, many field trials have provided the service for free or at artificially low prices to induce participation. Research about online services has demonstrated that there is a danger in attempting to extrapolate demand data collected when the service is provided free or at very low experimental prices if the goal is to forecast future demand at realistic prices (Flowerdew, Thomas, & Whitehead, 1978). Noll (1992) also argued that it is very difficult for large corporations to conduct field trials objectively. It is so difficult to launch a field trial and so many careers are on the line that companies often want to hear only positive results.

For those with sufficient commitment and resources, field trials have many benefits. Most importantly, they provide a real-world setting to test consumer usage and investigate demand. They also provide a realistic setting to explore engineering issues and understand actual costs in providing a service. This knowledge can help in scaling up from a single-market or limited test to a full-market service. Field trials provide an opportunity to develop working relationships with partners and to explore new applications of the technology. Equally important, they provide a setting to discover unanticipated positive and negative effects. The former can be built upon in a full-market launch; the latter can be addressed and potentially overcome before launching a full-market service.

FORECASTING PRACTICES AND ALTERNATIVES

The practical question arises: Should anyone try to forecast demand for new information services and, if so, what is the best way to approach it? Further, what are the alternatives to forecasting demand?

Our review of forecasting techniques and the history of forecasting new information services for the home suggests that current forecasting methodologies are weak both at a theoretical level and in general practice. Although it is difficult to overcome the theoretical limitations, forecasting groups can at least employ sound professional practices. This involves a clear understanding of the strengths and weaknesses of the research methodology employed, clear language about the service(s) that are encompassed by the forecast, and an explicit statement about the key assumptions that underlie a forecast. Further, key assumptions should be linked to the forecast in all subsequent reports about findings. It is also important to validate measures of consumer demand for the service. For example, do respondents understand the product or service they are questioned about and are responses an accurate indicator of future purchase or usage behavior?

In our general formulation of the forecasting problem, we emphasized the estimation of time necessary for a product or service to take off and have suggested it is better to build a conservative case about the time that is required. This may buy time to give the service a chance to become established and avoid premature withdrawal of the product from the marketplace when unrealistic expectations are not met. Unfortunately, the upfront requirements to win financial support for a new product or service often put pressure on sponsors and forecasters alike to develop a bias in the opposite direction and convey that the product will take off in an unrealistically short time frame.

Other analysts who have looked at the problem of forecasting methodologies have suggested that there may be a value in employing more than one methodology (Dimiru, 1984) and we agree. Multiple perspectives can provide a range in forecasts rather than a single estimate and highlight any differences in the assumptions that underlie each forecast. We have also suggested that there are many potential direct and indirect benefits associated with field trials and market trials, if a group has the resources (time, money, and personnel) for such an effort. It is important to emphasize that the potential direct benefit of achieving a more realistic demand forecast can occur only if a group understands that using a field trial to assess potential demand is a very challenging task and takes this objective very seriously. Often, a worthwhile exploration of demand is pushed aside by a desire to generate positive publicity.

It is also useful to ask why we engage in forecasting and whether there are any alternatives that could meet all or some of the objectives that underlie forecasting studies. Forecasting is a tool for business planning and, to a lesser degree, for policy development. It is intended to help organizations that create new services, investors in those services, and government agencies that must regulate them understand how a new service will develop in the marketplace. Typically, the forecasting effort is linked to a broader research effort that is intended to help identify who is most likely to adopt a service in its initial deployment and later, what are acceptable pricing levels for a service, what is the likelihood of failure to become established in the marketplace, what other technologies or services are likely to help or hinder establishment of the new service, and, when is the optimal time to launch a service, among other questions.

Forecasting techniques are quantitative tools to help assess how a new product or service will develop. There are also qualitative tools that can contribute to business plannning and policy development, as a complement or alternative to forecasting. They cannot provide statistical data about how many homes will use a new product or service 5 years after its introduction (nor, it may be argued, can forecasting methodologies do so reliably) but they can enlighten companies and policymakers about the process of diffusion as well as guide product development and implementation. They provide lessons and guidance for business planning by drawing upon an understanding of historical patterns in the development and diffusion of new information services. A few examples are reviewed here.

How Services Accelerate to Take-Off Speed. Marketing organizations often talk about “killer applications” or “magic bullets” that can lead to quick and decisive acceptance of a new technology or service. There are some examples of very popular applications that helped technologies to gain quick acceptance in millions of homes (e.g., a few very popular video games that drove the sale of video game consoles; Vogel, 1990).

However, it is more common for a new product or service to require the build up of several factors in order for the technology or service to take off and gain widespread acceptance. This build up of elements can occur relatively quickly or it can take many years. Broadcast AM radio and black-and-white TV provide examples of the former—they grew very rapidly after their introduction. Cable television and FM radio provide examples of the latter—they were in the marketplace for many years before they experienced a period of rapid marketplace growth (Sterling & Haight, 1978). The cable television and FM radio examples suggest that a technology or service can reach a threshold and then grow rapidly. The elements required to reach the threshold will not be the same for all technologies and the timetable for reaching the threshold can be a few years or many decades. Indeed, the crucial question is the time required to move from launch of a new technology to the threshold point for rapid growth. Further, many technologies never reach the point at which rapid growth becomes possible or they simply fail to gain marketplace acceptance.

Understanding these patterns, it is possible to analyze a new technology or service and assess at least some of the elements that must come together for the technology to take-off.

Early Versus Later Users and Uses. There is a broad generalization that applies to many technologies: The early uses and the early users for a technology may differ from later uses and later users. VCRs illustrate this process. When VCRs were first introduced in the United States, they were quite expensive. Early users included businesses and schools that used the technology for training and education as well as high-income households, especially those with an interest in the latest electronic gadget. Household usage included time-shift viewing of television programs (Dobrow, 1990) and a considerable amount of pornography (a majority of videocassettes sold and rented in the late 1970s was pornography). Businesses, schools, people who were willing to pay a high price for time-shift viewing of programs, and those who wanted to see pornography made up the initial group of users and uses. They made it possible for a second stage of adoption to occur, at a lower price and with a different mix of uses, including videocassette movie rentals and (later) videocassette sales.

In the case of VCRs, there were also some important unanticipated events. The emergence of "mom-and-pop" video rental shops was unanticipated and unplanned. Yet, these shops were critical for the second and third stages of VCR adoption to occur. This suggests that the growth of a technology is often a fragile, changing process. Early use can be different from later use and the elements that are critical to success at various steps along the way can sometimes come from unplanned and unanticipated sources.

The lesson here is that those who are introducing a new technology or service should try to anticipate the mix of users and uses at each stage. This is very difficult; thus they must be prepared to shift strategies as they move from early to later stages in a product's development.

Strategic Positioning: Early and Later Entry. A debate exists about the advantages and disadvantages associated with early entry of new information products and services. There are many examples of early market entry that escalated into market dominance. AM radio preceded FM into the marketplace and dominated radio for 50 years; HBO was the first to develop a national pay cable service and quickly dominated the market; and, the three broadcast networks that entered television in the late 1940s achieved a lock on the market that was not challenged for 30 years; among other examples.

However, for each example of early entry that led to marketplace dominance there is an example of early entry that led to failure or weak market performance. A 45 rpm automobile record player developed in the 1950s failed to achieve any significant market; two-way video trials and services in the 1970s for business meetings and medical applications were largely unsuccessful; and an over-the-air pay TV service developed by Zenith in the 1950s failed; among other examples. And yet each of these cases was followed by similar technologies and services that succeeded. There are many reasons why early market entrants fail. In some cases the technology simply doesn't work (e.g., the 45 rpm automobile record player skipped whenever the car hit a bump). In other cases, the costs associated with marketing and launching a service overwhelm an early entrant (e.g., several groups that planned to launch DBS services in the early 1980s abandoned their plans as they faced the huge costs associated with launching the services). In still other cases, an inhospitable regulatory climate can cripple an early entrant or consumers' lack of skill in using the new technology can lead to failure. Groups that follow early entrants may, in some cases, find that the technology works better, costs are lower, consumers have improved their skills in using the technology, the regulatory climate is more hospitable, and so on. Many argue that Japanese companies have employed the late entry strategy very successfully.

A historical review of new communication technologies suggests that early entry is an advantage in some cases and a disadvantage in others. It is an advantage when all the pieces are in place to launch the technology successfully. It is a disadvantage when the technology must stand on one or more Achilles' heels. Understanding this, a group developing a new information service can scrutinize it for weaknesses that could be crippling and try to overcome them or postpone a product launch until they are overcome. New service developers can also ask, "what elements must come into place for the product to gain acceptance in the marketplace?" If some critical elements are not in place, but are likely to emerge in a defined period of time, a late entry strategy may be advantageous.

Piggybacking on Replacement Cycles. The growth of some new technologies or services is linked to the purchase of other media. For example, few people buy a TV set or VCR just to obtain a picture-in-picture feature or stereo sound. However, when they purchase a new VCR or replace their old TV set many consumers select models with the latest enhancements such as picture-in-picture and stereo. In this sense, replacement cycles for existing media can provide an important way to introduce new products and services. It is reported that teletext in the United Kingdom grew in this way: People chose a teletext set when they leased or purchased a new TV.

Upgrade purchases have been very important for technologies such as television and personal computers where the pace of technological change has been rapid. The pace of change has been much slower in other technologies such as radio. In the case of telephony, the pace of change for end-user products and services was slow through the 1970s; it accelerated during the 1980s and 1990s.

If a new service can piggyback on purchases of other media, a developer can analyze replacement cycles for the host technology on which the new service will piggyback and the reasons why people replace existing technology in the home (e.g., to replace an old device that doesn't work or works poorly, to buy a unit for another room in the house, to buy an enhanced unit that can run the latest applications, etc.) to inform marketing efforts.

Lessons From Failure. There are also many lessons to be derived from technologies and services that failed in the marketplace or lost ground after achieving a significant penetration of households. These are often ignored as service developers concentrate on examples of success. For example, many technologies and services have failed because they offered a superficial benefit. Quadraphonic sound provides an example of a technology that failed for this reason. From a consumer's point of view, quadraphonic sound offered no advantage over existing stereophonic sound. It may be argued that electronic home banking to date has failed because it too has provided no benefits that consumers find meaningful.

In addition, technologies often fail when several competing standards confuse consumers or discourage manufacturers from bringing the technology to the marketplace. Teletext and AM stereo provide useful examples. Teletext flourished in the United Kingdom, where a single standard was adopted. In the United States, several standards competed for marketplace acceptance in the early 1980s. The Federal Communications Commission (FCC) failed to adopt a single standard and potential service providers disagreed about which standard to adopt. In this context, manufacturers were reluctant to build decoders, consumers were confused about what the technology offered, and teletext never emerged as a mass market service. AM stereo experienced a similar fate. Another lesson relates not so much to outright failure in the marketplace but to a decline in use of a service or technology after an initial period of growth. Subscription television (STV) and multichannel distribution service (MDS) illustrate the point. STV and MDS emerged in the 1970s as alternatives to cable TV by providing a form of over-the-air wireless cable service. STV and MDS grew steadily from the mid-1970s to the early 1980s, peaking in 1982 with approximately 2 million subscribers. They then declined through the 1980s as regular cable service expanded into more markets and consumers became disgruntled with poor reception from the wireless cable alternatives (Carey & Moss, 1985). However, some would argue that recent advances in microwave technology will put MDS in the category described next.

Failure in the marketplace may not be failure at all but an initial false start. For example, both television and VCRs experienced false starts. Television was launched in the late 1930s, but the high price of TV sets ($600) and disruption caused by World War II led to a false start. The technology was reintroduced after World War II and grew rapidly. Similarly, two home video-recording technologies were launched and then withdrawn in the early 1970s (the EVR system by CBS and Avco's Cartrivision system) before the modern VCR finally took hold in the mid-1970s.

The history of marketplace failures provides many lessons for the development of new information services. Those who ignore the lessons increase the likelihood that they will repeat earlier mistakes.

DISCUSSION

Zangwill (1993) argued that forecasting along with other forms of market research preceding the introduction of innovative new products and services is a "lousy idea." He cited Sony executive Kozo Ohsone—"When you introduce products that have never been invented before, what good is market research?" Zangwill suggested that, in the place of market research, companies introduce products into the marketplace, quickly measure consumer responses, and adjust the design of the product or withdraw it based on this feedback. This approach has more merit in the case of stand-alone products such as a Sony Walkman® where it is possible to exercise significant control over the manufacturing and marketing processes. It is more difficult in the case of a product or service with many hardware and software components, a very large initial investment and where there is less flexibility to respond to market feedback after the product is introduced (e.g., interactive television).

Given the poor track record of consumer demand forecasts, there is a certain appeal to casting aside all forms of research preceding market entry for a new service. However, this would sacrifice the important lessons that can be drawn from historical studies of earlier information technologies and services as well as the potential value of forecasting as a planning tool. In the case of forecasting demand, the key issues in deriving value from these planning tools are to understand the limitations of existing forecasting methodologies and to use professional practices. Historical studies of earlier information technologies can serve as a complement or alternative to forecasting but they cannot be used to create predictive instruments. Rather, they provide analytic tools to help identify components in the diffusion process. They can help to identify what differences will make a difference in consumer acceptance of new information products and services.

In addition, it is important for policymakers who regulate new information services, journalists who report about them and companies that create them to develop a realistic perspective about forecasting. For policymakers, this means focusing on those elements that are under their control that can foster the development of desired services. Forecasts have relatively little value in the policymaking process. Elements that are more useful include creating or encouraging plasticity in the technological infrastructure so as to make it easier, quicker, and cheaper for new applications to be tried out in real markets; reducing regulatory barriers that inhibit service development; and providing rate regulations that encourage groups to develop services and provide flexibility to adjust service offerings after they have been introduced.

For journalists, there is a need to understand how forecasting is done as well as the limitations of forecasting methodologies so that they can critically examine forecasts that come across their desks. It also means developing standards about the interpretation and use of forecasts as well as checking the reliability of sources for forecasts. In addition, journalists need to follow-up "press release forecasts" with hard questions about who did the forecast, what methods were used and what assumptions were built into the forecast.

For hardware and software companies, it is important to understand the fundamental limitations of forecasting methodologies and, where possible, to move away from a reliance on forecasts. As an alternative, we have suggested an analytic understanding of the diffusion process derived from studying the introduction of earlier and contemporary information products to the public. It is also important to position the introduction of a product strategically so that there is sufficient time (runway) to build critical mass (take-off speed) for acceptance of the product.

Commercial organizations may be reluctant to accept the limitations of forecasting methodologies or to rely on soft analytic tools in place of hard data. This would force them to confront a cold fact of life in bringing new information services to the public: risk. Most new products and services that are introduced to consumers fail. There is no reason to believe that new information products will enjoy a kinder fate. Indeed, as information services converge with the entertainment industry (e.g., in multimedia products) risk and uncertainty are likely to increase. There is a positive as well as a negative side to risk and uncertainty. If we cannot forecast the outcome for a new information product or service, it also means that its fate is written less in the stars and more in the actions of those who create, manage, and market the product or service.

REFERENCES

Arthur, B. (1990, February). Positive feedback in the economy. *Scientific American,* pp. 92–99.

Ben-Akiva, M. (1990). Choice of telecommunications services. In P. G. Holmlov (Ed.), *Telecommunications use and uses* (pp. 53–71). Stockholm: Teldok.

Brooks, J. (1975). *Telephone: The first hundred years.* New York: Harper & Row.

Carey, J. (1993). Looking back to the future: How communication technologies enter American households. In E. Dennis & J. Pavlik (Eds.), *Demystifying media technology* (pp. 32–39). Mountain View, CA: Mayfield.

Carey, J., and Moss, M. (1985, June). The diffusion of new communication technologies. *Telecommunications Policy,* pp. 145–158.

Dimiru, A. (1984). *An examination and critique of forecasting new technology trends.* Sussex: University of Sussex Department of Operations Research.

Dobrow, J. (Ed.). (1990). *Social and cultural aspects of VCR use.* Hillsdale, NJ: Lawrence Erlbaum Associates.

Dholakia, N., Bakke, J. W., & Dholakia, R. R. (1991, October). *Institutional patterns of information technology diffusion.* Kingston: University of Rhode Island Research Institute For Telecommunications and Information Marketing Working Paper Series.

Elton, M., & Carey, J. (1980). *Implementing interactive telecommunication services.* New York: The Alternate Media Center.

Flowerdew, A. D. J., Thomas, J. J., & Whitehead, C. M. E. (1978). *Problems in forecasting the price and demand for on-line information services.* London: London School of Economics.

Hough, R. (1979). *Pilot study to develop a methodology to forecast canadian demand for new home and business telecommunications services in the period 1980–1990.* Ottawa: Canadian Government Department of Communications.

Hough, R. (1980). *A study to forecast the demand for Telidon services over the next ten years.* Ottawa: Department of Communications.

Klopfenstein, B. (1989). Forecasting the adoption of new media. In J. Salvaggio & J. Bryant (Eds.), *Media use in the information age* (pp. 21–41). Hillsdale, NJ: Lawrence Erlbaum Associates.

Linstone, H. A., & Turoff, M. (1975). *The Delphi method: Techniques and applications.* Reading, MA: Addison-Wesley.

Lohr, S. (1994, February 19). New media face a clash of tastes. *The New York Times,* p. 37.

MacDonald, R. (1993, May). Technology Seminar at the Freedom Forum Media Studies Center of DBS forecasts by Wasserstein Perella Securities.

Martino, J. P. (1985, March). Looking ahead with confidence. *IEEE Spectrum,* pp. 76–81.

McCombs, M., & Eyal, C. (1980). Spending on mass media. *Journal of Communication, 30*(1), 153–158.

Noll, A. M. (1992). Anatomy of a failure: Picturephone revisited. *Telecommunications Policy, 16*(4), 307–316.

Sterling, C,. & Haight, T. (1978). *The mass media: Aspen Institute guide to communication industry trends.* New York: Praeger Publishers.

Northern Telecom. (1992, December 3). *Research shows 1990s may become decade of home information services.* [Press release]. Canada: Author.

Smith, R. L. (1970, May 18). The wired nation. *The Nation,* pp. 582–606.

Thompson, J. S., & Bowie, N. (1986). *Videotex and the mass audience.* Cambridge, MA: MIT Future of the Mass Audience Project.

Tyler, R. M. (1978). User research and demand research: What's the use? In M. C. J. Elton, D. W. Conrath, & W. A. Lucas (Eds.), *Evaluating new telecommunications services.* London: Plenum Press.

Vogel, H. (1990). *Entertainment industry economics.* Cambridge: Cambridge University Press.

Zangwill, W. (1993, March 8). When customer research is a lousy idea. *The Wall Street Journal,* p. A-12.

CHAPTER 4
Developing Strategies for Broadband Services

Paul M. Orme
Orme Associates, Ridgefield, CT

> The cable companies excepted, the new information road-builders agree on one thing: the initial market for all this "bandwidth" is not the much-touted one of the couch potato flipping through 500 channels of interactive television, punching buttons to get videos on demand, to buy goods from video catalogs or to play video games with viewers across the country. The market will emerge one day—and will eventually be huge. But the biggest group of users to ride the super-highway in the next five years will come from big business. That is because the information superhighway being built is not a freeway but a toll-road.
>
> —"America's Information Highway" (1994, p. 35)

The breakup of AT&T in 1984 unleashed a host of changes that marked the beginning of a period of rapid change. Local telephone companies, long distance carriers, cable television (CATV), and cellular operators are major participants in the telecommunications market in the 1990s; but their roles are changing and their appetite for other markets and a whole host of new activities is great. With telephone lines as a base, additional services such as data, graphics, and video are currently available by way of traditional broadcasting, coaxial cable, and satellites. The domestic telecommunications market is now estimated to be between $180 and $200 billion, from communications (telephone, cable, cellular, broadcast, and wireless), entertainment, video advertising, and online services such as Prodigy, CompuServe, America Online, Genie, Delphi and others.

The future promises much more—interactive, information, communication, transaction, entertainment, and shopping services, to name some. The Clinton administration is supporting the idea of creating an electronic "superhighway," available to a vast audience. These new services will require new means of delivery.

This continuously developing environment offers a multitude of new business opportunities. To take advantage of them, telephone, cable, information, and entertainment companies are reassessing their markets and developing new strategies. The first section of this chapter examines the supply-side pressures to offer broadband services to the residential market. Although the demand is not as yet sufficient to warrant offering these services, further analysis suggests that specific markets can be found for interactive broadband services. These markets are described in the last section and suggestions for marketing strategies are offered.

SUPPLY-SIDE PRESSURES

Telephone and cable companies must seek new markets for growth. Telephone companies, such as the seven Regional Bell Operating Companies (RBOCs) and GTE, are looking for new products, services, and businesses in which to invest and grow. The cable industry would seem to have peaked, not only in the extent of the market penetration but also in the rates for both basic and premium services; they also carry large debt, so financing for growth must come from outside investors. With telephone companies moving into video and entertainment services and cable companies moving into telephony, the competition is heating up between telephone and cable companies. Competition is likely to intensify and will be fueled by the new technologies such as direct broadcast satellite (DBS), wireless cable, fiber optics, and video compression.

Growing From Voice Communication. Telecommunications were originally limited to voice telephony. As the computer was adopted and as modems made possible transmission of data at modest rates of speed, the need to transmit both voice and data services over telephone lines arose. Telelphone companies provided these newer services, using narrowband technology to switch two-way voice and data communications, just as they had with traditional voice services. In the future, the new interactive services—information, communications, transactions, entertainment, and shopping services—require a switched broadband data and video network, and the deployment of new technologies. Telephone companies have the option of adding new sophisticated network technology to the existing twisted-pair wire system or creating an entirely new broadband network with the new fiber optic technology. Telephone companies are seeking to expand their previously limited service; and move beyond their revenues of more than $107 billion in 1991. Local telephone companies account for a large share of that revenue and even in long distance service, the seven RBOCs are important because 70% of the volume and 50% of the long distance revenue are generated within their own territories, where the local telephone company must cooperate with long distance companies.

In the mid-1990s, total revenues from telecommunications include both local and long distance telephone, plus CATV, as well as associated advertising over cable and broadcast, videocassette rental and purchase, which together added another $70 billion to the $107 billion (just noted) in 1991 alone. The telephone companies continue to dominate this business, accounting for more than 65% of total revenue. As they strengthen their existing services and develop new ones, they will likely face competition in the local telephone market, and some analysts believe that profit margins may be reduced by as much as two thirds.

A major competitor is likely to be CATV. Although CATV has expanded from 35 to 75 channels in some markets, the tree and branch structure of its system limits it to one-way delivery. Interactive services will require a new method of delivery. Cable companies are also seeking to enter the local telephone business to compete directly with telephone companies and both wireless and wireline cable sponsored local telephone service tests have been announced.

Targeting Existing Video Entertainment Markets. The opportunities to create services by adding interactive expanded video programming has placed cable companies, telcos, and entertainment program providers in a race for dominance in interactive television, and has been fueled by almost daily press attention. Cable companies have broadband delivery technology as well as access to the entertainment and video programming, known in the industry as "content." Telephone companies have neither the broadband delivery technology nor the content, but they do have experience in switching voice and data services.

The new participants in the market such as DBS and wireless cable operators are initially targeting the existing video entertainment business (CATV, pay per view, videocassette rental, etc.). Some of these new entrants may offer interactive services as well. They may be able to enter the market at any time and offer higher quality digital television (and more than 100 channels!), at a lower cost. Regulation also works to their advantage because the Cable Act of 1992 guarantees them access to programming. In addition, because wireless cable is not governed by the Cable Act, operators are free to compete on price and has more flexibility in making decisions.

The existing video markets will be the target of both the cable and telephone industries as they attempt to develop interactive television and video-on-demand. Competition will reduce revenue to each. The telephone industry may be at a disadvantage because of its monopolistic past and lack of marketing and strategic planning experience. Cable companies have experienced problems with customer service and negative reaction to price changes. However, both cable and telephone companies also have the advantages of already being in the marketplace and being able to offer a complete package of services, perhaps at lower cost.

Strategic Alliances. The new technologies and the services they make possible have caused massive shifts and realignments in the affected industries, making alliances advisable for a stronger position in the changing marketplace. In addition to strategic alliances, companies are making acquisitions, creating investments, and deploying technologies in response to their vision of the future. The 1993–1994 avalanche of announcements about mergers, investments, and partnering from telephone and cable companies sounded more like monopolists seeking the comfort and protection of big partners than a response to market demand. In a sense, the discussion has been supply-side driven; since we have the technology, lets use it and assume the market for it will follow. If we build it, they will come, to quote *Field of Dreams.*

How quickly will interactive television and video-on-demand grow? Will there be sufficient revenue and profits to justify the major investment in new technology? Telephone and cable companies may spend as much as $1,000 per household to make fiber and video compression available. A result of this may be a significant reduction in the cost of telephone service and it may be possible to allocate to telephony much of the cost of conversion. Because of these considerable costs, however, even a significant share of the entertainment market will not be sufficient reward. It is possible that offering the video-on-demand and video telephone may simply produce tough competition and transform the industry into a commodity business:

> Many in the industry believe the Baby Bells' monopolistic upbringing has led them to underestimate the effect of competition on prices. Pessimists reckon that carrying basic services, such as video-on-demand and videophone calls, will swiftly turn into a low-margin commodity business. . . . This would be bad news for an industry expected to invest an estimated $20 billion a year in multimedia networks (and even more on programming) during the next decade. ("Multimedia's Yellow Brick Road," 1993, p. 67)

WILL THERE BE A DEMAND FOR BROADBAND SERVICES?

The success of this new business will be determined by consumer demand for broadband and interactive services. Many industy participants believe that offerings that traditionally have been popular—television shows, movies, and games—should be joined by shopping and video conferencing services. In 1991, consumer purchases of entertainment services and products (basic cable subscription, video rental, recorded music, movie and theatrical box office, and miscellaneous amusements) was estimated at $45 billion, or $475 per household, per year (Bilotti, Hanson, & MacDonald,

1993). Will these new interactive services increase the overall market or simply draw from existing markets, such as videocassette rental and premium cable channels?

Initial Indications of Demand

Tangential to the entertainment market is a number of related ones that are likely targets of the interactive services industry. These include education, admissions, consumer electronic equipment, and reading materials (see Table 4.1). The size of these markets, totaling more than $2,000 on average, per household, per year, makes them commercially very attractive indeed.

Consumer Interest in Interactive Services. The chairman of Bell Atlantic, Raymond Smith, predicted the first services for interactive television will be movies on demand, interactive games, increasingly sophisticated home shopping, and direct response advertising. He made no estimate of anticipated revenues, but in late 1993 he expressed the belief that the business will be significant.

Examining Smith's projections in more detail, the following are the markets that Bell Atlantic sees as targets:

Movies on Demand: To compete with the $12–$15 billion videocassette rental/sales business.

Interactive Games: To compete with the $6 billion in-home video games business.

Sophisticated Home Shopping: To compete with home shopping from QVC and the Home Shopping Network, which achieved $2.5 billion in 1993; and the $60–$70 billion catalog shopping market.

TABLE 4.1
Annual Household Expenditures

Category	*Estimated Annual Expenditure (Per U.S. Household)*
Telephone services	$617.93
Education	$446.66
Admission fees (movie, theater, opera, sporting events, and other)	$377.83
TV, radio, & audio equipment	$287.76
Cable TV (includes noncable homes)	$180.08
Reading materials	$163.03
Total	$2073.29

Source. The Official Guide to Household Spending (1994).

Direct Response Advertising: To compete with direct marketing and telemarketing; direct mail advertising was $27 billion in 1993, and total advertising expenditure reached approximately $138 billion in 1993.

According to one research firm, Americans are unlikely to spend significantly more for entertainment because average household income is down by $2,000 in real dollars from its 1989 high. Affluent households are therefore the likely target market for more entertainment and video services, but that is the group that seems to have the least leisure time.

In March 1993, Backer Spielvogel Bates, the advertising agency, conducted a survey of 1,010 adults. When asked about their interest in interactive television, about 32% said they would subscribe. But certain population segments that participated in this survey had an above average interest in subscribing (see Table 4.2).

In a 1993 CBS/*New York Times* survey of 1,347 adults, respondents were asked about their interest in interactive TV. Replays of past shows, or one form of video-on-demand, received a positive response: 77% of respondents would like to have ready access to replays of favorite shows, but 51% said they had no interest in ordering products or paying bills via interactive television.

Willingness to Pay. Consumers are spending money to interact with various forms of information services, home TV shopping and/or hand-held interactive games. The revenue from these activities is approximately $9–$9.5 billion.

Online services such as Compuserve, Prodigy, America Online, Genie, and Delphi had $750 million in 1993 revenues and approximately 4 million users. America Online has announced plans to expand to devices other than the personal computer.

TABLE 4.2
Interest in Interactive TV

Subjects (N = 1,010)	*Interest*
Ages 18–34	48% higher than average
Incomes	
$20,000–$30,000	37% higher than average
$40,000 plus	16% higher than average
Househholds	
with children	35% higher than average
with three or more members	25% higher than average
People who went to college	19% higher than average
People in the Western U.S.	11% higher than average

Source. Backer Spielvogel Bates Agency, 1993.

Home shopping, offered by the Home Shopping Network and QVC, totaled approximately $2.5 billion in revenues in 1993. The audience demographics are very different from subscribers to online services or interactive games.

Interactive games account for $6 billion in revenue per year, and more than one third (and some industry observers estimate an even higher level of 42%) of U.S. households have an interactive game player.

This comparison should not, however, be seen as predictive of interest in future interactive services. Research by a CBS/*New York Times* poll suggests that consumers are not willing to pay very much for interactive TV. When asked "How much would you pay per month?" only 47% indicated willingness to spend between $5 and $10 and 48% expressed willingness to spend more than $10 per month (see Table 4.3).

Experiences From New Service Trials. Several trials are currently underway in various parts of the country and these trials have generated data regarding consumer demand for interactive services. Since 1991, GTE has tested interactive and video-on-demand services in Cerritos, California. Fewer than 5% of cable customers were willing to pay $9.95 per month for interactive TV services that allowed them to shop, pay bills, check stock market quotes and news, play electronic games, make plane reservations, take sample Scholastic Aptitude Tests, and scan a video and audio version of the *World Book Encyclopedia.*

The second GTE service, a video-on-demand service called "Center Screen," offered a 30-channel, pay-per-view (PPV) system featuring movies available within 30 minutes and costing $3.95 to $4.95 per movie. This video-on-demand service achieved an "ever used" incidence of just 57%, well below the 77% (noted earlier) interest expressed when price was not mentioned. Feedback from some respondents who have lived with the

TABLE 4.3
Willingness to Pay for Interactive TV

Amount Per Month	*Percent of Respondents*
Nothing	5%
Less than $5	22%
Less than $10	25%
Subtotal	52%
Between $10 and $15	20%
Between $15 and $25	19%
More than $25	9%
Total	100%

Source. CBS/*New York Times* Poll, March 1993.

Cerritos video-on-demand system suggests that they would rather rent movies from video stores because it is less expensive.

In July 1992, AT&T, Tele-Communications, and U.S. West began a test of video-on-demand and near video-on-demand among 320 households in Littleton, Colorado. The test was named Viewers Choice TV and offered 24 channels of staggered-start movies-on-demand to one group of households. To another group, total video-on-demand was provided via banks of VCRs. Test customers were not charged for installation or service. The results were that 95% of orders for video-on-demand were placed shortly before use—similar to PPV experience; 2.5 movies were rented per month (vs. 2.6 PPV per year estimated by Paul Kagan Associates), or 12 times more; consumers were paying 99¢ to $3.99 per movie and payment per month averaged between $6 and $7 (in line with research quoted earlier); and about 70% of the households used these services per month versus a PPV statistic of approximately 20%.

The ultimate demand is uncertain. Results from various surveys and test markets do not suggest a convincing consumer appetite for the potential 500 additional channels that technology could provide. Consumers are unikely to pay a premium for interactive services and video-on-demand. The estimated monthly charge of $7.50 for interactive services and a share of the videocassette rental market will not be sufficient to warrant the huge costs. Ultimately, the video entertainment market may not provide sufficient benefit to justify the cost of interactive television. Specifically, revenue and share of market taken from consumer-direct purchases of entertainment and advertising from traditional television, direct marketing, and print media may not provide enough of a market for interactive services to justify the cost of investment. "The biggest question-mark over multimedia is whether the market will be as big as the Baby Bells, cable companies and folk in Hollywood believe" ("Multimedia's Yellow Brick Road," 1993, p. 68).

As John Sculley, former chairman of Apple Computer, asked, "Are we expecting too big a behaviour change, too quickly, by everybody from programme creators to viewers?" ("Multimedia's Yellow Brick Road," 1993, p. 68).

NEW GROWTH OPPORTUNITIES

Traditional wisdom in consumer-goods marketing theory would preclude even test marketing a product or service with an average interest to subscribe of only 32%. On the other hand, if this 32% is compared to the 60% of households that already subscribe to CATV, one could say that slightly more than half of those who pay for CATV have an interest in interactive TV. Despite the interest in the typical consumer in-home market, it seems that interactive and broadband services have significantly better business opportuniteis in areas that have not received press coverage to the extent that entertainment, video-on-demand and interactive games have.

Initial demand for broadband interactive services to the home may come from those for whom the services will have an economic or productivity payoff. The work-at-home market—more than 20 million individuals in the mid-1990s—and the telecommuter market—already large and continuing to grow—are likely to have a need for video telephone and data transmission services, including graphics, fax, data transfer, messaging and numerous other productivity-support services. Work-at-home and telecommuters represent individuals who work from their home or other nonoffice locations. Work-at-home and telecommuters have adopted new technologies at a much higher rate than the population as a whole and are likely to take advantage of interactive and broadband services. They may well lead the way for recreational markets.

Home education and home health care (HHC) monitoring/telemetry are other markets where interactive video and television services, including video telephony, may be welcome, and provide sufficient revenue to warrant the investment that is estimated to reach tens or even hundreds of billions of dollars. Yet another possible market is shopping and the ordering of merchandise. While QVC and the Home Shopping Network have grown to approximately $2.5 billion in annual sales, many in the industry believe that the market can grow well beyond this level into that of catalogue and retail sales. Finally, gaming/gambling may be another form of entertainment with a future on interactive and broadband networks. Gaming/gambling has grown rapidly over the last decade to become a $30 billion industry. With imagination, the new technologies could add significantly to the growth of this business. Each of these business opportunity areas are discussed here.

Work at Home

The number of people who work at home either as self- or company-employed increased to 30.4 million in 1992, up 9% from the previous year. Another 8.6 million company employees worked from home outside of normal business hours and used telecommunications or electronic equipment support such as personal computers, modems, fax machines, or an extra telephone line. Home workers in increasing numbers are dependent on computer and telecommunications technology to handle key office tasks, to provide the image and competitive capability of a larger organization, and to improve productivity.

There are several kinds of at-home workers. They are detailed in Table 4.4.

TABLE 4.4
Work-at-Home Market

Type	*Amount*
Telecommuters	6.6 million
Self-employed	12.1 million
Moonlighters (part-time workers)	11.7 million
Subtotal	30.4 million
Corporate (Bring work home)	8.6 million
Total	39.0 million

Source. Link Resources (1992).

Dollar value of home office products and telephone services by all categories of home workers is estimated to be in the range of $26 billion—an increase of nearly 20% in 1992 versus 1991. It is divided as follows: $12 billion on work-related phone calls; $5.5 billion on PCs and other hardware; $3 billion on other electronics for home; $2 billion on software; $1.5 billion on online services; $1 billion on peripherals; $1 billion on fax devices for a total of $26 billion (see Orme, 1994).

Telecommuters

Company employees who work at home are known as *telecommuters.* In 1992, there was an estimated 6.6 million of them, and they used telephone, facsimile, computer, and other electronic links to connect them to the office or to other business participants. Unlike other work-at-home categories, the telecommuter segment is supported: Products and services are paid for by employers. An employee may wish to become a telecommuter for reasons of lifestyle, but is unable to do so until his or her employer provides the infrastructure and funding.

There are several forces that appear to be encouraging telecommuting. The Clean Air Act of 1990 requires employers to develop methods of reducing commuting trips, so the number of telecommuters is likely to grow. Companies may have additional financial incentives to adopt telecommuting if technology can meet their needs. For example, they may be able to dramatic savings on real estate costs. Productivity is likely to be increased as time in meetings is not wasted. Finally, companies might find highly talented personnel who, for a number of reasons, might otherwise not be available.

Although current technologies may offer narrowband communications, fiber, compression and wireless technologies provide the basis for more complete broadband interactive communication. With wider use of these new technologies, more companies may be willing to adopt telecommuting, even for sophisticated work practices. The ability to hold video conference calls from virtually any location, including desk tops in workers homes, will offer a dramatic additional dimension to telecommuting and will all but reproduce the face-to-face office environment. As more broadband technology is introduced and sophisticated services are used by telecommuters, more companies, even those with complicated communication needs, will find telecommuting an advantage.

Telecommuter categories are expected to grow rapidly in the next decade with increasing societal acceptance and use of technology. In addition, telecommuting will become more cost effective as broadband technology becomes more widely available.

Education

Education may be one of the largest areas for interactive and broadband services as increasingly more schools adopt computers and off-premises learning programs. Home education programs will provide interactive voice, data, and video services among parents, students, teachers, counselors, administrators, librarians, and other education professionals. Service capabilities will expand as technologies advance. Such services will be a gateway to a wide array of external sources and databases, and may be placed into four categories:

1. Archival information (e.g., videotapes, library materials, online informational databases),
2. Live (e.g., lectures, classrooms, labs, videoseminars/conferences),
3. Electronic bulletin boards/E-mail/voice mail (e.g., homework assignments, daily class coverage, test results, event calendars, meeting schedules, messages among students/parents/teachers/librarians/administrators), and
4. Broadcast (e.g., live or taped programs from various educational sources).

These services could maintain a complete educational profile for each student and make individualized recommendations from computer-based expert systems software, linking home education services (HES) with computer-assisted instruction (CAI). Key information could be downloaded from the database to a "smartcard" (EduCard™) that would allow users entry to HES gateways from different locations/devices and automatically bill user accounts.

In addition to traditional educational sites, distance education (DE) will play a central role in the reform initiatives, both because of DE's potential to deliver cost-effective improvements in education and its synergies with current education trends. In addition to DE, there are several trends in education that will work well with HES and will therefore nurture it. These include greater parental involvement in education, growth of the home-educated student population, growth of adult education, privatization of public schools, integration of "special needs" students into the mainstream, increasing shortage of trained librarians, greater business community involvement in education, and wider student population diversity.

As noted later, BellSouth has made a commitment to support the linking of education and health care in a nine-state area. Homes and facilities such as schools, hospitals, clinics, and laboratories are likely to be affected. With interactive and broadband capabilities coming into the home, the opportunity to deliver educational products and services and health care support and monitoring systems will be significant.

A recent announcement by BellSouth (1994), as reported by Reuters:

> The BellSouth Corporation introduced a fiber optic telecommunications network today [January 26, 1994] that will initially connect public service sites, including schools, medical centers and science operations in North Carolina."
>
> Describing the system as an "information superhighway," BellSouth, the largest regional telephone company, said the venture was part of a new breed of high-speed multimedia networks planned for it's nine-state territory in the Southeast.
>
> 'This is a realization of an interconnected network, which allows people to interact with one another and with institutions and culminates years of planning,' said Duane Ackerman, president of BellSouth Telecommunications phone service unit.
>
> The system will connect 52 schools, 16 universities, 18 community colleges, 11 medical centers, 4 state government sites, 3 public safety offices, and 2 scientific sites. The networks will be connected by more than 116,000 miles of fiber optic cable, BellSouth said.

Home Health Care

Home health care monitoring/telemetry will begin in narrowband applications leading to interactive video, television services, video telephony, and other new applications of technology for the HHC market. In the future, the market for HHC services is expected to grow dramatically.

The U.S. Department of Commerce estimated expenditures of $10.7 billion for HHC services alone in 1992, up from $4.1 billion in 1987, a 5-year compound annual growth rate of 21.1%. There are several drivers contributing to the growth of this industry, including changes in the demographics and lifestyles of the population, cost-containment efforts by the government, and insurance payments and professional expectations.

Demography and Lifestyle—Prelude to the Boomers. The elderly population is the HHC industry's main market—accounting for 69.2% of all patients, and 68.9% of all visits in 1990. According to the U.S. Bureau of the Census, between 1990 and 2000 the elderly population will grow rapidly, especially the 75+ age group who require the most care: The 55+ age group will grow 11.6% to nearly 60 million people; and the 75+ age group will grow by 26.2% to nearly 17 million people.

Life expectancies continue to increase—from about 79 years in 1991 to 82 years in 2000. As mortality rates decline, the number of people with chronic conditions (e.g., diabetes, hypertension, heart disease, etc.) requiring long-term HHC will actually increase because medical technology keeps them live longer. The growing affluence of the elderly means there are more health consumers in homes suited to HHC more insurance coverage to pay for it, and more people able to afford high quality HHC, which is less expensive than institutional care. The demand for health care will be fueled by the baby boomers; in 2002 the oldest "baby boomers" (47 years of age in 1993) will just be under a decade away from "senior citizenship." When this enormous age cohort reaches the age ranges associated with increasing needs for medical care, the nation's health care delivery system will be severely stretched.

Government and Insurance Payments—Cost Containment. A clear effort is underway to control health care costs because HHC is dramatically less expensive than hospital care, government and private reimbursements were revised to shorten hospital stays, resulting in increased HHC. From 1964 to 1988, hospital inpatient days decreased 14%, whereas outpatient days grew 35%; in 1990, 40% of surgical procedures were conducted on an outpatient basis. This trend will continue as technology support and acceptance of HHC care becomes more commonplace.

Medical Profession—Bedside Manner Back in Vogue. Home-based care is seen by the medical profession as generally less traumatic for patients than institutional care. The position of the AMA Council on Scientific Affairs regarding HHC is:

> Home care is a rapidly growing field that is beginning to attract greater physician interest and participation. . . . Medical care in the home is highly diversified and innovative. The areas of preventive, diagnostic, therapeutic, rehabilitative, and long-term maintenance care are all well represented as physicians develop new practice patterns in home care. . . . Home care is an increasingly important aspect of medical practice. . . . As practicing physicians become more involved with home care, we can expect to see even greater innovation and growth in this new/old field of medical practice. (Council on Scientific Affairs, 1990, p. 1241)

Home Shopping

Analysts believe that TV-based home shopping/ordering businesses will seek to grow from the current $2.5 billion by taking market share from the $60 billion catalogue business and the $250 billion retail-department store business. Consumers currently purchase products and services from home using the telephone and direct mail. In fact, nearly half of the population has made purchases from catalogues. One survey showed that consumers bought an estimated $60 billion from catalogues in recent years and nearly half of all households made catalogue purchases (see Table 4.5).

TV home shopping customers have been categorized as junk-jewelry-buying women, but recent research has shown a very different picture. In the past, studies had shown men to be only 20% of the TV shopping audience, but QVC has tried to change this by offering more products to the male audience. One of the problems of TV home shopping, in general, is the high level of returns, which are reported to be as high as 50%.

In a related area of home shopping via PC/modem, the top three product areas are travel, home electronics/computers, and banking/investment/insurance. When asked which types of electronic TV shopping consumers would like, books and travel-related services topped the list (see Table 4.6).

Gaming/Gambling

Interactive television will provide a means to deliver gaming in new ways and may produce an expanded market. Immediacy combined with the spontaneity of gaming shows may provide the opportunity to develop interactive TV gambling as a new, exciting activity.

TABLE 4.5
Consumer Catalogue Purchases

All Households	*46.6%*
Clothing	57.3%
Home furnishings	18.4
Housewares	15.6
Toys/games	13.4
Sporting goods	10.2
Electronics	8.2
Food gifts	7.5
Gardening	5.9
Hardware	5.8
Food	3.6

Source. Simmons Research on *Media and Markets, Annual National Survey* (1992).

TABLE 4.6
Consumer Interest in Electronic Shopping*

Product	*% Agreeing*
Books/CDs/tapes	40%
Hotels	34
Air travel	31
Vacation destinations	25
Women's apparel	21
Telephone services	13
Financial transaction products	12
Financial loans	9
Investment products	9
Savings products	9

Note. (*N* = 1,220 adults with annual income of $25,000 or more)
*Includes screen phones, PCs, and TV-based systems.
Source. Coopers & Lybrand (1993).

Legalized gambling grossed a record $329.9 billion in 1992. Gambling gross revenues have nearly tripled in 10 years. After winnings were paid, $29.9 billion was available for governments and gaming establishments—more than six times what people spent on movie tickets. Legalized gambling of one form or another operates in every state of the union except Utah and Hawaii. Some analysts believe that gambling has been legalized as a way for politicians to raise money without increassing taxes; gaming is seen to be an easy escape for them (see Table 4.7).

CONCLUSION

For both telephone and cable companies, it is likely to be many years before the interactive superhighway is functioning, and it probably wont be until

TABLE 4.7
Revenues from Gaming/Gambling

Category	*1982 (billions)*	*1987 (billions)*	*1992 (billions)*
Lotteries	$2.2	$6.6	$11.5
Casinos	$4.2	$6.4	$10.1
Pari-mutuel	$2.8	$3.4	$3.7
Bingo	$0.8	$0.9	$1.1
Indian gaming	NA	$0.1	$1.5
Other	$0.4	$0.9	$2
Total	$10.4	$18.3	$29.9

Source. Orme (1994).

well into the first decade of the next century that it will be widely available. However, competition between telephone and companies will increase and new technologies and industry segments will emerge.

The competitive video recreation and entertainment market are not likely to be large enough to support the significant investment to deploy interactive broadband support technologies. As a result, other market opportunities must be sought. To survive and prosper in this competitive environment, new strategies must be developed to cater to those segments of the home market that respond to economic and productivity rewards. These appear to be the work-at-home segments that have adopted to the new technologies at a much higher rate than the population as a whole; the HHC segment that is being driven by cost-containment pressures and demographic changes; and the education segment. In addition, home shopping and gaming/gambling appear to be two more applications with attractive revenue potential that can benefit from the excitement of interactive and broadband services.

REFERENCES

America's information highway: A hitch-hikers guide. (1994, January 7). *The Economist,* pp. 35–38.

Backer Spielvogel Bates Agency. (1993, March). *Future effects of new consumer and commercial communications technologies. New York: BSB Media Research & Technology Department.*

Bilotti, R., Jr., Hanson, D., & MacDonald, R. J. (1993, December 8). *The cable television industry. New technologies, new opportunities and new competition: Vol. I: Industry review and outlook.* New York: Grandchester Securities, a division of Wasserstein Perella Securities.

CBS/*New York Times* Poll. (1993, March) as reported in the *New York Times.*

Coopers & Lybrand (1993, January). *Electronic Shopping Study.* Presented at *Interactive* Services Association (ISA) Conference. Washington, DC.

Council on Scientific Affairs. (1990). Home care in the 1990s. *Journal of the American Medical Association, 263*(9), 1241.

Link Resources. (1992). *Work at home market.* New York: Author.

Multimedia's yellow brick road. (1993, December 4). *The Economist,* pp. 67–68.

Official Guide to Household Spending Report. (1994, January 31). *The Marketing Pulse Newsletter,* Woodstock, New York.

Orme, P. M. (1994). *Interactive and broadband strategy development in a changing communications environment.* Ridgefield, CT: Paul M. Orme & Associations.

Reuters. (1994, January 25). BellSouth fiber optic plan. *New York Times,* Sec. D., p. 5.

Simmons Research. (1992). *Media and markets: Annual national survey.* New York: Author.

CHAPTER 5
Taking Movies-on-Demand to Market

Ruby Roy Dholakia
The University of Rhode Island

> Movies may not turn out to be the most magical kind of content in this new medium . . . content will earn the richest profits in the new TV industry . . . and the one that network owners are betting on is movies and TV shows at the touch of a button.
>
> —"Feeling for the Future" (1994, p. 10)

From the time of the early Greek theater, enactment of human and mythical stories has been a major form of entertainment for large numbers of people. Actors and troupes traveled from village to village bringing the stories and dramas to people dispersed over the countryside. With technological advances, the diversity of forums and variety of entertainment continue to expand audiences the world over. First the silent movies and then the talkies allowed the presentation of performances without live actors. The large, flickering silver screens attracted thousands to movie theaters everywhere. Radio, and later the small screen TV, gradually shifted the entertainment settings from theaters and auditoriums outside the home to the "theater of the mind" and the intimacy of individual livingrooms. This shift has intensified with cable TV (CATV) and videocassette recorders (VCRs). As a result, we find increasing numbers of households gathered before their TV sets at home. Since the 1950s, the household hours spent watching TV have steadily climbed, reaching 49.5 hours per week in the 1990s (Papazian, 1994).

Changing technologies and changing consumer behaviors have attracted the attention of many suppliers and there are many efforts to increase entertainment choices for the U.S. consumer. In this chapter, we examine the efforts currently underway to bring movies-on-demand to the U.S. consumer. From concept development to limited market trials, we examine the supply-side pressures to find a killer application for broadband services for the residential consumer. The possibilities of success are examined through a study on consumer interest in movies-on-demand and implications are drawn for taking this new broadband service to market.

VIDEO ENTERTAINMENT AT HOME

Until 1975 when HBO introduced commercial-free movies, CATV had difficulty gaining consumer acceptance; by 1985, CATV had become a $20 billion industry (Piirto, 1993). VCRs accelerated the viewing of movies at home with more than 77% of VCR viewing time spent on prerecorded material and almost all of it for movies—new releases, catalogue movies, children's features, classic movies, and adult movies (RAB, 1994). In 1991, consumer spending on film entertainment exceeded $31.6 billion and nearly 85% of that was spent on film entertainment at home (Graves, 1993).

In the mid-1990s, movies have become the prime form of video entertainment—whether through free TV, pay cable TV, or VCR. As Table 5.1 indicates, the largest share—nearly 85% in 1991—of dollars spent on film entertainment is spent in the home. When we examine the time spent, we find an average of nearly 12 hours per week spent viewing movies in the home, the single largest category of TV set usage in the home. Although broadcast TV accounts for a significant share of that time, the "free" access to entertainment is reflected in its zero share of entertainment expenditures.

Structuring Video Choices in the Marketplace

The allure of movies, inspired by Hollywood, continues to spin magic and memories for U.S. viewers as well as for audiences worldwide. Today, the average American can meet his or her movie viewing needs in a variety of ways. Choices are extensive, accessible, and affordable. In addition to attributes such as genres and actors, there is tremendous choice in terms of availability—place, price, and time. A recent survey of movie viewing indicated that more than 75% of consumers saw movies at home twice or more a month (K. Carter, 1992). Movies are also seen at theaters but not as frequently; of those seeing movies at a theater, only 30% saw two or more movies during the month (K. Carter, 1992). By 1993, 38% of the U.S. population received one or more movie channels, and more than a third of

TABLE 5.1
Share of Consumer Expenditure (Time and Money) on Film Entertainment

Movie Delivery Medium	*Expenditures*[a]	*Viewing Time*[b]
Outside home in theaters	15.2%	NA
Inside home		Weekly hours = 11.93 (100%)
Home video	34.8%	18%
Basic cable	35.1%	18%
Pay cable	14.9%	19%
Broadcast TV	0%	45%

Sources. [a]Graves (1993). [b]Papazian (1994).

these homes received one or more pay-per-view (PPV) channels. An average of 1.3 PPV movies was seen in 6 months in 1993 (Story, 1993).

The increased emphasis on entertainment at home has created opportunities for a large number of businesses in addition to the traditional TV networks and movie studios, including video rental and sales outlets, various CATV services such as HBO, Showtime, as well as PPV movies. Besides specialty video stores, video rentals are available in supermarkets and convenience stores.

To structure consumer choice, movie releases are staggered by offering movies first in theaters, then as PPV, next as video rentals and finally as network TV (Lyle & McLeod, 1993). Although there was a trend toward a shorter window between releases in different formats, Universal Pictures at least, is moving back to 60-day windows for PPV movies after trying a 30-day window; Disney, Columbia, and 20th Century Fox are continuing with 30-day windows ("Online News," 1994). Similarly, the pricing of movies affects the locations as well as frequency of movie viewing. The cheapest movies are available on video: The average price for a one-night video rental is $2.50 compared to $4.89 for a movie theatre ticket (*RAB,* 1994). There is also an increasing trend toward video purchases as the release price of prerecorded movies keeps declining. This is particularly true for repetitively viewed movies such as children's features.

The changing dynamics in the structure of supply and demand generate excitement and challenge in meeting the consumer need for video entertainment. Between 1990 and 1993, the average number of videos rented has declined; and surveys suggest that increasing percentage of customers fail to rent at all when faced with the unavailability of their first choice. Analysts have implied "consumers are taking their dollars away from video rentals completely and spending that money on a different form of entertainment" (Story, 1993, p. 23). Between 1988 and 1993, revenues from PPV films estimated to have been grown from about $50 million to $170 million (Brown, 1993). Cable TV providers are enthusiastic that "movies will lure mainstream consumers into other types of interactivity" (Piirto, 1993).

MARKETING VIDEO-ON-DEMAND MOVIES

Movies-on-Demand: The Killer Application?

The increase in entertainment time and dollars spent at home is continuing to attract the attention of existing and new players. As Brown noted (1993), more than half of the 20.8 million cable households with access to PPV are not being well marketed. Telephone companies are interested in getting a share of the increased video revenues (Kupfer, 1993). Near-video-on-demand and video-on-demand (VOD) applications are being trialed in various parts of the country. Every single Regional Bell Operating Company (RBOC) as well as CATV and media entertainment companies have announced technology and market trials of video-on-demand ("Broadband Trial," 1994; Mundorf & Dholakia 1994). In their search for "killer applications" to bring the information superhighway to the U.S. home, movies and shopping appear to be the two major applications on which the marketers are willing to place their early bets (Allen, Ebeling, & Scott, 1994; "New Yardstick," 1994). According to Kaplan (1992): "Video-on-Demand (VOD) may be the first giant step in making interactive media part of the American lifestyle" (p. 43).

From Consumer Need to Service Concept

Several studies have supported overall consumer interest in the concept of video-on-demand (Allen, Heltai, Koenig, Snow, & Watson, 1993; Cermak, 1993). Using a simulation, Allen et al., for example, reported significant consumer interest in enhanced PPV as well as movies-on-demand. Supported by such evidence, several versions of video-on-demand and near-video-on-demand are being formulated by telephone companies, CATV companies, and other entertainment houses (Mundorf & Dholakia, 1994). In each of these cases, the attempt is to offer convenience of at-home viewing with near instant availability of first choice movies at prices comparable to video rentals.

Most of these research reports are proprietary and access to details of the results are sketchy and limited. In order to understand more about consumer reactions to VOD, a study was designed to examine the relationship between consumer characteristics and interest in VOD movies.

EXPLAINING CONSUMER INTEREST IN MOVIES-ON-DEMAND

A Research Study

As part of a research project on technology in the household, a survey was conducted among households in Southeastern New England during 1993–1994. The households were part of a mail panel on technologies in the home. Approximately 500 households responded to the multiple waves of questionnaires and provided information on several new product and service concepts.

One new service concept was movies-on-demand, given as a short description (see Exhibit 1). Although the concept was based on the use of

Exhibit 1: Movies-on-Demand Concept

> **TELE CINEMA:** With Tele Cinema (also called VIDEO-ON-DEMAND) one can request a movie on the TV set at any time using the telephone. Tele Cinema enables viewing of any movie at any time without leaving the home. You simply dial the toll-free number and Tele Cinema sends the movie over the phone lines to your TV set. It does not require the use of the VCR. But like the VCR, one can stop the movie and continue again at a later time. Like premium cable TV movie channels, there is a one-time installation fee for Tele Cinema. After installation, there is a charge for each movie viewed.

phone lines requiring no special set-top device, the main purpose of the study was to determine the relationships between consumer characteristics and interest in the concept. A structured questionnaire was used to measure the perceptions of the concept as well as the willingness to adopt the concept.

Research Framework

The research has been influenced by the following framework (see Fig. 5.1). Three sets of consumer variables are key to understanding how the concept of movies-on-demand is perceived: demographic characteristics of the consumer (e.g. gender, age, education, income, and family size), current movie viewing behaviors (at home and in theaters), and current investment in video technologies (e.g., TV sets, VCRs, video cameras, cable movie channels, and video rental memberships).

Demographic variables influence TV and video use behaviors generally. For regular TV use (broadcast as well as basic CATV), available evidence suggest that females watch more TV than males and older consumers watch more TV than younger consumers (Papazian, 1994). When video use is expanded to include the use of premium CATV channels, PPV channels, and VCR, data from a variety of sources suggest that heavier video users are mostly male, mostly middle-aged, and mostly from upper income groups. Gender, age, and income also influence investment in equipment and services such as VCRs, camcorders, videotapes, cable TV subscriptions, and video rentals (e.g., Times Mirror Center, 1994).

Movie-viewing behaviors represent established patterns of behavior that can be sustained or curtailed by new services such as movies-on-demand. Because of the positioning as in-home entertainment, movies-on-demand should appeal to people who are already viewing movies at home through the use of videotapes and/or CATV movies. Extension of the same logic suggests that people who currently view movies in the theater should find movies-on-demand less appealing than people who view movies at home.

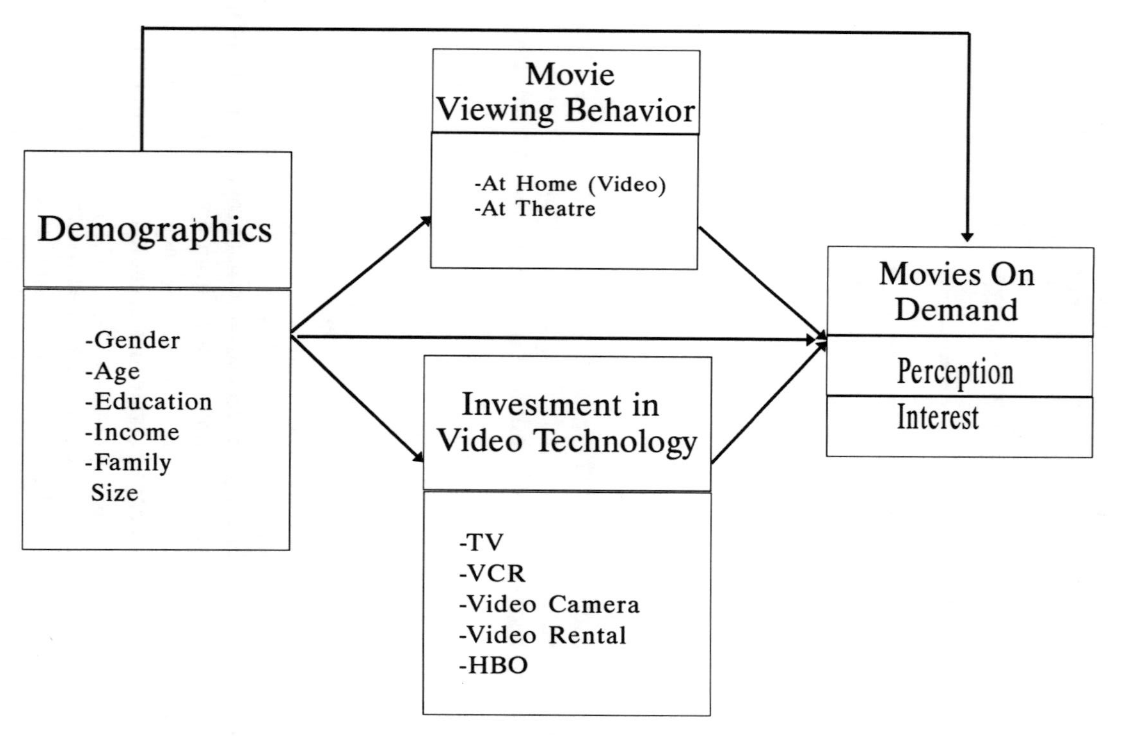

FIG. 5.1. Influences on movies-on-demand.

Investment in video technology is the final characteristic of interest in this research study. Investment in technology through ownership of equipment and subscription to services suggests consumer involvement in the product category. A consumer who has invested a great deal in available video technologies offers behavioral proof of interest in the product category. This consumer is more likely to be interested in new offers within that category. To the extent prior investment in the category is made obsolete or less useful by new introductions, we can expect lower consumer interest in the new service based on relative advantage. On the other hand, if the new service expands or facilitates the use of existing products/services, then we expect consumers to show greater interest in the new service. Although movies-on-demand require the use of existing TV and telephone connections, they can also negatively affect the usefulness of VCRs and other video products and services. We need to understand how prior investment in video technologies affects the perception of and interest in movies-on-demand.

Consumer Interest in Tele Cinema

Interest in adopting the new concept was measured by three statements averaged to create one score ranging from 1 (*not interested*) to 5 (*very interested*). Overall interest in the concept of Tele Cinema is modest (average = 3.2, sd = 1.1, n = 490). There are, however, a significant segment within the survey respondents (25%) who express strong interest in the concept of Tele Cinema and indicate agreement to seeing a demonstration of the new service, to trying out the new service in their homes, and to purchasing the service. Another 20%, however, was definitely not interested in the concept, whereas the rest appeared to show only moderate interest.

In order to explore the nature and significance of differences in the interest among various demographic and behavioral groups, a multiple regression analysis was performed. The results are summarized in Table 5.2. Taking all the variables together, the overall model is able to explain 50% of the variance in consumer interest in Tele Cinema.

It is the perception of the concept as "fitting with the consumer's lifestyle" (i.e., perception as "likely to be used often," as "popular with friends," as "more exciting than renting from video stores," as "a real time saver") that contributes the most to consumer interest in Tele Cinema. Furthermore, investment in video technologies as well as movie viewing behaviors contribute positively to the interest. Specifically, people who view movies—whether at home or at the theater—are interested in Tele Cinema but the interest is greater among those who view video movies at home than in the theater. Individuals who have invested in technologies such as VCR, video

TABLE 5.2
Explaining Consumer Interest in Tele Cinema*

	Consumer Interest in Tele Cinema
Variables included in overall model	R^2 = .50, F = 42.2, df = 10/424, $p < .00$
Demographics	Age[a] – negative; Gender[b] – negative
Movie viewing behaviors	Movie at theater[a] – positive Video movies at home[a] – positive
Investment in video Technologies	Investment in video techs.[a] – positive
Perception of Tele Cinema	Fit with Lifestyle[a] – positive

*Only significant relationships reported. [a]Regression coefficient significant at $p < .001$. [b]Regression coefficient significant at $p < .05$.

rental memberships, HBO, TV sets, also show a positive interest in Tele Cinema. Finally, only age has a statistically significant and negative relationship among the demographic variables. Interest is higher among younger groups. The highest degree of interest is expressed by 18 to 24-year-olds and the lowest degree of interest is expressed by those over 65. Gender has a modest relationship with interest higher among males than females.

Explaining "Fit With Lifestyle"

The empirical evidence generated in this specific study suggests that the Tele Cinema concept, if perceived to be fitting with lifestyle, will be of interest to potential consumers. Perception of Tele Cinema as fitting with lifestyle is greater among those consumers who have made commitments to this form of entertainment (Table 5.3). Those who view video at home perceived the Tele Cinema as fitting more with one's lifestyle than those who view

TABLE 5.3
Explaining Tele Cinema as "Fitting With Lifestyle"*

Explanatory Variables	*Perception of Tele Cinema as "Fitting with Lifestyle"*
Variables included in model	R^2 = .12, F = 7.7, df = 8/445, $p < .00$
Demographics	Age[a] – negative Education[c] – positive
Movie viewing behaviors	Video at home[a] – positive Movies at theater[b] – negative
Investment in video technologies	Investment in video techs.[b] – positive

*Only significant relationships reported. [a]Regression coefficient significant at $p < .001$. [b]Regression coefficient significant at $p < .05$. [c]Regression coefficient significant at $p < .10$.

movies at the theater. In fact, those who saw more movies at the theater viewed the concept negatively in terms of fit with lifestyle. Investment in related technologies such as VCR, video camera, video rental memberships is also positively related to the perception of the Tele Cinema concept as fitting with lifestyle. Among demographic variables, age is most important and is negatively associated with the perception of Tele Cinema as fitting with lifestyle, whereas education has a marginal but positive relationship. These variables together explain 12% of the variance in the perception of the concept of Tele Cinema as fitting with lifestyle.

The findings from the research study suggest that movies-on-demand concepts such as Tele Cinema is not likely to have universal appeal. Although the overall interest is quite modest, a significant segment of the surveyed consumers show interest and these are consumers who perceive the concept to fit their lifestyle. This fit is highest among younger and more educated consumers, among those who currently view movies at home and have invested considerably in video technologies.

MARKETING IMPLICATIONS FOR VOD MOVIES

To succeed, VOD service concepts will have to appeal to selected market segments. This appeal is highest among those consumers who currently view movies at home and have invested in technologies that allow viewing movies at home. Although the cost and service implications of attributes such as variety, scheduling intervals, pause and rewind, and so forth, are immense, the success of any movie-on-demand offer will depend on its ability to match the convenience, flexibility, and economics of current video viewing behaviors.

This is supported by other studies. In a laboratory experiment to estimate demand for VOD movies, Cermak (1993) found video rentals to be central in the allocation of video dollars and closely related to all other sources of video entertainment. Higher prices of VOD movies, for instance, created a increased demand for video rentals; higher video rental prices, however, did not have as strong an effect on VOD purchases.

From Concept to VOD Trials

Currently, there are several companies engaged in test marketing VOD. Some estimate the number of trials to be 50 or more (Elmer-Dewitt, 1994; Ziegler, 1994). Local telephone companies such as SNET, GTE, US WEST, Bell Atlantic, Rochester Telephone Corp., long distance and network systems companies such as AT&T and Sprint, as well as entertainment companies such as Time Warner and Viacom are all involved with testing

VOD or near-VOD ("Broadband Trial," 1994; Mundorf & Dholakia 1994). In the Quantum trial in Queens, New York, 50 of the 150 channels will offer dozens of movies that will start every half hour (Piirto, 1993).

Many of these are technology trials utilizing a combination of fiber and coaxial cables to deliver the video signals to the household. Others are market trials attempting to determine the customer features that are attractive as well as the revenue potential from such services. Several of these trials start on a modest scale, frequently using a "captive" user group to assess the technological features. Some have brought users into the "laboratory" to assess the features and their attractiveness whereas still others have taken the technologies to the users' homes and used a more realistic setting to determine market acceptance of VOD. Most trials, of course, have followed all these methods over many years to develop and deploy VOD.

There are several issues that will affect the successful commercialization of VOD (Auletta, 1994). Obviously, for the corporations involved with test marketing VOD, there are enormous outlays of monies (Higgins, 1994). For example, SNET has estimated that its costs will be $4.5 billion over the next 15 years to enhance Connecticut's telecommunications systems (Davies, 1994). Ameritech plans to spend $4.4 billion to install fibre directly to the homes and businesses of its customers (Nolan, 1994). The cost of the set-top device is a critical issue: Time Warner is testing a Silicon Graphics workstation priced at more than $3,000 while waiting for a more affordable device; Viacom is testing a set-top box costing one tenth that amount (Rebello, Hof, Weber, & Mallory, 1994). It will be very difficult to generate revenues to recover these costs in any reasonable period of time. Payback periods have been revised by Ameritech based on its experiences in trial video dial tone projects in Illinois, Michigan, and Ohio (Stern, 1994). The maturing of the CATV industry and its financial limits have also been raised as a concern (B. Carter, 1994). Many companies are engaged in partnerships to help defray the associated costs ("The Baby Bells' Painful Adolescence," 1992; Lambert, 1994). Vela Research and C-Cube Microsystems have agreed, for example, to jointly develop VOD systems based on their areas of mutual expertise (Vela Research, 1994).

Technical complexities remain to be resolved. One set of problems is created by the immense quantity of "bits" of information required to store and forward each movie. Technologies like Asymmetrical Digital Subscriber Line (ADSL) are being considered capable of providing "VCR-quality movies on demand, competing with video rental stores and cable pay-per-view, but offering a greater selection and superior customer control" (Fleming & McLaughlin, 1993, p. 20). Companies like AT&T, Microsoft, and Oracle are racing to develop "servers" that will be able to retrieve, assemble, route and deliver customer requested movies (B. Carter, 1994).

A second set of problems is created by the likelihood of large numbers of customers demanding the same movie at a specific time. "Compressing and storing video is only half the battle. Making sure customers can quicky and accurately order the movie of their choice is equally important. Vela draws on HSN's enormous experience with order and billing services to complete the vital conncection betwen the customer and the service" (Vela Research, 1994).

Consumer interest in more frequently scheduled PPV movies in the Quantum project in New York does not seem to be a solution favored by many operators, many of whom feel convenience is less important than variety of choice (Umstead, 1994). Others are trying to offer convenience. In Denver, for instance, a new feature will enable viewers to take a break from a movie and pick it up 15 minutes later on a different channel (*Atlanta Journal Constitution,* 1994).

Some analysts believe that true VOD will not be possible and compromises will have to be made. "It won't make sense to fetch and feed a hit movie 1,000 times for 1,000 customers. Better to go for the lower-cost 'virtual video on demand.' Start the movie every minute or five minutes, then feed it to everyone who has requested the film since the previous start" (Burgess, 1994, p. F1). Some service providers are attempting to offer different levels of service and distinguish those customers who want to pay for true VOD. SNET, for instance, in its Hartford, Connecticut trial, is planning to offer "Reel One" as enhanced PPV with a lower price, whereas "Command Performance" is offered as VOD at a higher price (Mundorf & Dholakia, 1994).

Others are attempting to offer services beyond movies-on-demand. NYNEX, for example, is offering movies-on-demand, news-on-demand, as well as sports-on-demand (Brown, 1994). "First Choice TV" hopes to provide missed episodes of TV shows on demand (Jessell, 1993).

Many issues still remain unresolved. Browsing behavior is an important influence on video rentals. Interest in home delivery of rentals was not supported by willingness to pay a higher price for the convenience; instead researchers found that customers felt they could not use it because "they would not know what to rent without being in the store to see what's available" ("Multimedia: Movie Theatres," 1994).

The allocation of entertainment dollars is a major issue (Auletta, 1994). Some evidence seems to suggest that the total entertainment dollars has increased in recent years, particularly at the expense of household savings (Mandel et al., 1994). Between 1979 and 1993, spending on entertainment and recreation as percent of all non medical consumer spending increased from 7.71% to 9.43% (Mandel et al., 1994). Earlier research seems to suggest that entertainment dollars are a fixed proportion (McCombs & Eyal, 1980) and only with rising incomes can one expect absolute dollar expenditures to increase. Incomes, however, have not been rising in recent years. Although surveys indicate that consumers are willing to spend some money on interactive services, the amount of additional expenses is uncertain (Allen et al., 1994; Find/SVP, 1994; Jessell, 1993). Many of the VOD trials involve a period in which the service is provided for free. Upon reaching a certain point in time, the customer is then charged for the service provided. SNET will allow its customers to sign up free for 3 months and then charge the customers the remaining 9 months of the project (Leder, 1993). Subscription behavior involving actual expenditures of consumer dollars will be an important test that still needs to be completed.

"Quantum's biggest accomplishment is that its customers are buying PPV at eight times the rate of traditional systems. In some months, the buy rate averages 1.5 movies per subscribing household" according to H. Panero, former vice president of marketing (cited in Piirto, 1993). TCI reports 2.5 movies a month at VOD test trials in Colorado (Elmer-Dewitt, 1994). The price–cost relationships in these trials are not clearly known. Others are less optimistic. "Right now pay-per-view is just the electronic version of the home video. The only advantage is that you don't have to return the tape. It is doubtful that it will break through to the mainstream until it proves the convenience of stop-and-start viewing. To win the war, it must become video-on-demand" to quote Gerbrant (cited in Piirto, 1993, p. 38).

CONCLUSION

Over several decades, the U.S. consumer has invested in video technologies, spending increasing amounts of time and money on entertainment at home. As a result, there has been great interest on the part of various suppliers to structure consumer choices in order to attract and retain a share of this growing marketing. Video entertainments are now available in various forms, prices, and places.

Movies-on-demand is being offered as one more option for entertainment at home and various trials are currently underway to test technologies and market responses. A research on the concept of movies-on-demand shows there is moderate interest in the concept, and that this interest is greatest among younger age groups. The experiences with the trials indicate that the widespread availability of movies-on-demand is still several years away as technological complexities continue to challenge the deployment of this service. Market accepance is also a problem at this stage. Although interest in movies-on-demand is higher among those who view movie videos at home and have invested in video technologies, the positioning of movies-on-demand is also a challenge because of its acceptance by this segment. Current competitive alternatives pose technical and economic hurdles that movies-on-demand must overcome.

It is very clear that suppliers will continue to push the deployment of movies-on-demand as one of the first applications for bringing the information superhighway to the home consumer. The wide availability of content and the extensive experience in offering the content over multiple media provides a platform on which movies-on-demand can be built. The providers of this new service can focus on technical and consumer issues to appropriately fine tune the service offering with consumer demand. It is also very clear that movies-on-demand will only be one of several applications that will ultimately support the costly building of the superhighway infrastructure. Managers are hopeful of offsetting these immense costs through cost cutting measures and through the creation of increased demand. As one was quoted to say, "VOD systems are so expensive, you wouldn't do it just to watch movies. In the long run, the major source of revenues and profits will be from transactions (such as shopping) and advertising" (Cook, 1993, pp. 63, 66).

REFERENCES

Allen, D. L., Jr., Ebeling, H. W., Jr., & Scott, L. W. (1994). *Perspectives on the convergence of communications, information, retailing and entertainment.* Washington, DC: Deloitt Touche Tohmatsu International.

Allen, J. R., Heltai, B. L., Koenig, A. H., Snow, D. F., & Watson, J.R. (1993, January/February). VCTV: A video-on-demand market test. *AT&T Technical Journal.*

Auletta, K. (1994, April 11). The magic box. *New Yorker,* pp. 40–45.

Broadband trial players and their equipment suppliers. (1994, June/July). *On Demand,* pp. 12–13.

Brown, R. (1993, November 15). Study shows PPV revenue up in 1993. *Broadcasting & Cable,* p. 35.

Brown, R. (1994, May 2). Video on demand: What price freedom? *Broadcasting & Cable,* pp. 11.

Burgess, J. (1994, January 19). Computers get a blockbuster role. Firms race to develop "servers" providing millions with movies on demand. *Washington Post Financial,* PF1.

The baby bells' painful adolescence. (1992, October 5). *Business Week Special Report,* pp. 124–134.

Carter, B. (1994, May 23). After years of easy profits, cable industry hits slump. *New York Times News Service.*

Carter, K. (1992, February 10). USA snapshots: A look at statistics that shape our lives. *USA TODAY.*

Cermak, G. W. (1993). *Budget allocation as a measure of potential demand* (Working Paper). Waltham, MA: GTE Labs.

Cook, W. J. (1993, December 6). This is not your father's television. *U.S. News & World Report,* pp. 63,66.

Davies, P. (1994, January 16). Network vital to survival for SNET. *New Haven Register,* pp. F-1, 2.
Denton, C. (1993, December 6). Pioneering the electronic frontier. *U.S. News and World Report,* pp. 56–63.
Elmer-Dewitt, P. (1994, May 23). Play . . Fast . . Forward . . Rewind . . Pause. *Time,* pp. 44–46.
Feeling for the future: A survey of TV. (1994, February 12). *The Economist,* pp. 1–8.
Fleming, S., & McLaughlin, M. B. (1993, July 12). ADSL: The on-ramp to the information highway. *Telephony,* pp. 20–26.
Find/SVP. (1994). U.S. consumers are willing to spend $5 billion for interactive information access. [News release]. New York: Author.
Graves, T. (1993, March 11). Industry surveys: Leisure time. *Standard and Poor's.*
Higgins, J. M. (1994, June 27). Are telcos really this rich? *Multichannel News,* pp. 6-8A.
Jessell, H. A. (1993, April 26). John Hendricks's voyage of discovery. *Broadcasting & Cable,* pp. 39–41.
Kaplan, R. (1992, June). Video on demand. *American Demographics,* pp. 38–43.
Kupfer, A. (1993, January 25). Any movie, anytime. *Fortune,* p. 83.
Lambert, P. (1994, June 13). Playing well with others. *Multichannel News,* pp. 8A–11A.
Leder, M. (1993, October 22). Videos on demand to undergo pilot test. *New Haven Register,* pp. 25, 29.
Lyle, J., & McLeod, D. (1993). *Communication, media and change.* Moutain View, CA: Mayfield.
Mandel, M. J., Landler, M., Grover, R., DeGeorge, G., Weber, J., & Rebello, K (1994, March 14). The entertainment economy. *Business Week,* pp. 58–64.
McCombs, M. E., & Eyal, C. H. (1980). Spending on mass media. *Journal of Communication, 30,* 153–158.
Multimedia: Movie theatres, home video. (1994, March 21). *Dow Jones News Service.*
Mundorf, N., & Dholakia, R. R. (1994) Video on demand in the United States: A survey of trials and market potential. Kingston: RITIM, College of Business Administration, The University of Rhode Island.
New yardstick for interactive. (1994, May 23). *Broadcasting & Cable,* p. 6.
Nolan, C. (1994, June 6). Ameritech calling. *Cablevision,* p. 28.
Online news. (1994, June/July). *On Demand,* p. 7.
Papazian, E. (1994). *TV dimensions '93.* New York: Media Dynamics.
Piirto, R. (1993, May). Taming the TV beast. *American Demographics,* pp. 34–41.
Public Relations Announcement (1994). Vela Research and C-Cube Microsystems establish development partnership. (1994, March 24). [Public relations announcement].
Radio Advertising Bureau (1994, April). *Instant background. Video stores.* Washington, DC: Author.
Rebello, K., Hof, R. D., Weber, J., & Mallory, M. (1994, March 14). Interactive TV: Not ready for prime time. *Business Week,* pp. 30–31.
Story, B. (1993, December 5–11). Profile of the video customer: Consumer survey. *Video Store Magazine,* pp. 12–26.
Stern, C. (1994, May 16). Ameritech: More time before break-even. *Broadcasting & Cable,* p. 52.
Times Mirror Center For the People & the Press. (1994, May). *The role of technology in American life.* New York: Author.
Umstead, T. (1994, June 6). Operators not ready for 15-minute movies. *Cablevision,* p. 22.
Ziegler, B. (1994, May 18). Building the highway: New obstacles, new solutions, *Wall Street Journal,* p. B1.

PART II
User Perspectives

CHAPTER 6
Entertainment as the Driver of New Information Technology

Jennings Bryant
Curtis Love
University of Alabama

It is a truism that entertainment will be the driver that at long last creates adequate market demand for advanced information technologies in the home. This notion has become an article of faith among telecommunications service providers, who have launched trial after trial looking for entertainment-based "killer applications"[1] for new technologies and services.

The idea of a "trial"—pretesting new products and services in one or more test markets before offering them for widespread adoption—is an example of utilizing inductive reasoning (Bacon's logical induction, or deriving general principles from specific observations) to help make business decisions. To date, this model of inductive reasoning has been the most common approach used to determine whether killer applications for information and communication technologies exist and, if so, what they are.

In this chapter, we review and synthesize the findings from industry trials regarding the potential of entertainment to drive new communication technologies. Prior to that, however, we lay groundwork by reviewing some of the literature that forecasts the adoption and use of new information technologies, discusses barriers to the adoption of advanced information technologies, describes the special potential of entertainment applications for getting consumers to adopt new media, and looks at the shifts to the home as a promising market for information technology. Following the examination of industry trials, we focus on some of the special features of information and communication technologies that experts think will be the keys to making entertainment innovations drivers of the information utility of the future. We conclude by offering the challenge of a new approach to testing the entertainment potential of information technology in the home.

[1]The notion of a killer application is borrowed from the computer software industry. It refers to a computer program that catches on so strongly with consumers that it sells not only the software but hardware and peripherals alike (Stern, 1994).

INDUSTRY FORECASTS

> The role of information is transforming the nature of the economy. (Arrow, cited in Mandel, 1994, p. 22)

> The multi-billion dollar question . . . is whether consumers will share the industry's excitement about technology that transforms the living room TV set into a "digital interactive multimedia terminal." (Kehoe, 1994, p. 5)

Industry forecasts for the economic impact of information and communication technologies vary widely. For example, a recent article (Fatsis, 1994) admitted that "the big mystery is how much consumers will pay for interactive information and entertainment" (p. 8) and then cited long-range industry economic forecasts of $3.5 trillion per annum. Many experts note that the present contribution of information technologies to national or international economies is not a good indicator of their future potential because only now do we appear to be approaching the threshold of the era of profitable deployment of information technology in residential markets. Therefore, more immediate economic forecasts are modest in comparison. For example, having defined *information technology* broadly as any product and service that enables the creation, manipulation, and dissemination of information, Hamilton (1994) predicted information technology to be a $469 billion market worldwide in 1994.

According to U.S. Department of Commerce Secretary Ron Brown, the information industry in the United States alone already exceeds that figure. On January 6, 1994, Brown announced that the information industry is fast becoming the predominate factor for the U.S. economy. He reported annual U.S. revenue from the information services industry of $610 billion for 1993 and further estimated that over the next decade, more than one-half of all U.S. jobs will be in the information industry, particularly in the telecommunications sector. Brown also announced that, in recognition of the increased prominence of information technologies and services in the national economy, for the first time the Commerce Department had grouped a number of communications and information services into one independent category. This included computer software and hardware, information services, electronic communications components, telecommunications services and equipment, and search and navigational equipment ("Technology," 1994).

During 1993, the leading revenue gainer of the group was computer software, which was expected to achieve 12.8% growth in revenues in 1994 to $60 billion. Information services was projected to be the second largest gainer in 1994 (12.4%) and should realize revenues of $136 billion in 1994. The electronic information sector was projected to grow at 15% to $15.6

billion and should continue at or above the 15% growth mark for the next 5 years. Telecommunication services were projected to increase at 8%: Local exchange services should rise 3%, domestic long-distance telecommunications revenues are expected to increase 6%, and international long-distance services should jump by 20% or more in 1994. In the area of entertainment, the focus of this chapter, the forecasted growth rate was 6.1%, reaching $268.3 billion in spending by 1997.

Tempering these reports of rapid growth for information technologies and services are those urging caution regarding these optimistic forecasts. A report in *American Demographics* (MacEvoy, 1994) noted that "SRI analyses suggest that many growth projections for new media applications in entertainment and information services are too optimistic. They estimate that the universe of households using new media will range between 5 million and 30 million by 2010" (p. 42).

An *Adweek* article echoed these sentiments: "It may take 10 or more years for companies to deliver just the services that consumers are being told about today, to say nothing about newer, more advanced services" (Frankel, 1993, p. 28). Another sobering note posted in *Billboard* indicated that "the dawn of interactive [technologies is] not yet upon us and may not be for some time. . . . Consequently, even though new methods of electronic entertainment delivery are technologically possible today, it will be years before these methods are in widespread use" (Fishman, 1994, p. 4).

Is the glass half empty or half full? We may well be at a point in time where it is impossible to tell. If research on the adoption of other media tells us anything (e.g., Salvaggio & Bryant, 1989), it would appear that as far as information technologies are concerned, we are still at a relatively early portion of the S-shaped diffusion and adoption curve. This is especially true for entertainment services designed and targeted for the residential marketplace. At present, the "early adopters" are having their turn with trials, beta tests, and limited distribution rollouts. The responses of these samples of consumer opinion leaders to the services they are testing may well be key determinants of the gradient of the adoption curve to come. Literally millions of consumer responses will be formulated in the next 2 or 3 years, and they will undoubtedly hold the key to the veracity of the optimistic versus pessimistic forecasts. The reason for this uncertainty is simple. As many of the forecasts have noted, "this industry is in its infancy" (Lohr, 1994, p. A-1).

BARRIERS TO ADOPTION

> In six months I expect to see a huge backlash against all of this. (P. Andrews, 1994, p. D2)

> After a year in which the idea of a 500-channel world went from fantasy to commonplace to cliché, the backlash has begun. ("What If They're Right?," 1994, p. 3)

One of the factors that will help determine how consumers respond to the potential of advanced information technology is the way the media cover "the information technology story," which helps set the public agenda for the topic. The potential and problems of advanced information technologies and systems have been much in the news of late, with a substantial portion of the coverage focusing on the National Information Infrastructure (NII) or the so called information superhighway. A recent investigation by Bryant, Gonzenbach, and McCord (1994) provides a trend analysis of coverage of such information technology issues by six major newspapers.[2]

Figure 6.1 illustrates the number of articles devoted to information technology and the information superhighway per month between September 1992 and February 1994. As can be seen from examining this figure, the topic became "hot" during the fall and winter of 1993. Key events precipitating increases in coverage were the release on September 15, 1993, of a federal report entitled "The National Information Infrastructure: Agenda for Action," which unveiled the administration's NII initiative, and a December 15, 1993, speech at the National Press Club in Washington, DC by Vice President Al Gore, at which he announced that the NII would "help unleash an information revolution that will change forever the way people live, work, and interact with each other" (Bureau of National Affairs, 1994, p. N8).

More important than the quantity of coverage, however, is what is covered and how it is covered. Bryant, Gonzenbach, and McCord (1994) adapted Down's (1972; cf. Neuman, 1990) *issue-attention cycle*:

- Preissue stage: the issue or phenomenon exists, but it has not yet captured major press or public attention;
- Discovery stage: there is a sudden steep ascent of attention and emphasis on the potential or the problem;
- Plateau: there is a gradual realization that if the potential is real, the problems are complex and not very easy to solve;
- Decline: the press and public become inattentive and possibly frustrated with the problem; and
- Postissue stage: the issue enters a period of inattention, although its objective conditions have not necessarily changed significantly.

---

[2]The six papers were the *New York Times, Washington Post, Los Angeles Times, Dallas Morning Star, Chicago Tribune,* and *Atlanta Journal & Constitution.*

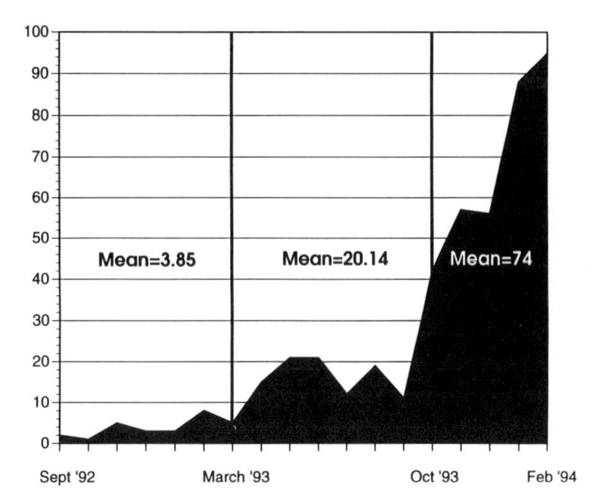

FIG. 6.1. Frequency of media coverage (September 1992–February 1994).

Bryant et al. found that most of the newspaper articles printed before December 1993 fit into the preissue stage, in that many product or service innovations of information technologies were presented and discussed, but the issue typically was identified as being in the domain of scientists, engineers, and the technology companies. Moreover, although the potential of information technologies and services was discussed quite often, little consideration was given to the complexity of the issues. In contrast, the trade press was already debating the potential and problems, with scholars and innovators clearly aware of the complexity of the issue. From December 1993 to January 1994, most articles seemed to represent the discovery stage. The NII initiative clearly captured press and public attention, and considerable excitement was generated. Beginning toward the middle of January 1994, however, the "tone" of many of the articles began to change, reflecting a lack of excitement, more recognition of the challenges that lay ahead, a focusing on the complexities of the issue, and quotes from more of the "nay-sayers" and those emphasizing barriers to adoption of new information technology.

Although no formal assessment of press coverage since February 1994 has been conducted, close scrutiny of recent coverage of information technology issues seems to reflect the plateau stage of a focus on barriers to adoption. A sampling of this press coverage from 1994 reveals the range of barriers discussed during this period:

> "This may be an aged, conservative, rear-guard position," says Howard Stringer, president of CBS, "but I know of no evidence that viewers are crying out for more television. Stringer's Law says that, as channels multiply, standards deteriorate."

Such positions might seem more aged, conservative and rearguard were they not shared by so many people with no financial stake in thwarting change. For Americans who can barely cope with the 35 TV channels (on average) they already have, the prospect of 465 more is understandably disconcerting. Scanning them would be a nightmare; engineers say it would take 43 minutes. And paying for such a system could be painful. As one woman told the *Wall Street Journal,* "I feel like laughing when they talk about 500 channels. Your cable bill would exceed your monthly mortgage payments. ("From Prime Time to My Time," 1994, p. 9)

Laws ensuring universal service (eventually) can militate against networks creating classes of information haves and have-nots. But with the least educated citizens often the most technophobic, and a tragic number functionally illiterate, such public policies may not be enough to fend off the inegalitarian effects of new communications technology. Worse, those effects may be magnified internationally, as the rich world is wired faster than the poor one. ("It's the End of the World as We Know It," 1994, p. 17)

While technologists, entrepreneurs, and deal makers work on everything from windows-based TVs to mega-mergers, no one has really asked what it is that this new technology is supposed to produce, or how it's supposed to produce it. (Dickinson, 1994, p. 63)

Last summer Pam Edstrom, Microsoft publicist, was discussing the media's sudden infatuation with interactive television and the "information superhighway." Mergers involving telephone, cable television and entertainment giants were just starting to dominate headlines. Concepts such as video on demand, virtual communities and digital repurposing were not yet part of everyday lexicon. But Edstrom already saw the writing on the wall. "In six months I expect to see a huge backlash against all of this," she predicted. (P. Andrews, 1994, p. D2)

Marketers of interactive television and other new media products and services may damage the very market they are trying to build, warns a new study released Thursday. "If companies initially miss the mark on what consumers are after, they risk locking out key segments of the population or delaying their willingness to try new media products," said Nick Donatiello, president and chief executive officer of Odyssey (the market research firm which conducted the study). ("Nationwide Study Reveals Risks," 1994, p. 1)

Much of this talk is (almost) as wild-eyed as anything that came out of McLuhan. The difference is that it is not loopy professors but the business studs themselves who are doing the talking. Listen to Ray Smith, boss of Bell Atlantic, an American telecoms company: "We stand on the verge of a great flowering of intellectual property, a true Renaissance that will unleash the creative energies of investors, entrepreneurs, hackers, artists and dreamers. . . . There is no shortage of doubters. After a year in which the idea of a 500-channel world went from fantasy to commonplace to cliché, the backlash has begun. . . . Not all the doubters have a vested interest in nostalgia; some are among the industry's boldest moguls. Rupert Murdoch reckons that bosses are "a bit over-excited by this digital deluge." Ted Turner agrees. "Every single interactive TV experiment has failed," he said. "Most people want to sit back and watch—interacting is hard work." ("What If They're Right," 1994, p. 3)

But there is a downside to all this magic. Coping with 500 channels will make the editing of which ones to scan in any one viewing session essential. And, of course, there is the electronic log of what you have been watching recently in the box's memory banks that allows the cable companies to "poll" their boxes and track who is watching what. . . . While you're viewing, Big Brother already is watching. (Coates, 1994, p. C10)

Admittedly, many of the issues raised in these and numerous other recent articles in the popular press—equity and parity, problems with managing information abundance, costs, excessive superhighway hype, privacy—really are potential barriers to widespread adoption of information technologies in the home. At the very least, these issues must be given ample attention, discussion, and research investigation if the social, economic, and political potential of advanced information technologies is to be realized. But the sensationalistic way the press often has handled such topics of late undoubtedly also contributes to a larger potential problem: lack of consumer confidence and a relatively uninformed questioning of whether the uses of advanced interactive technologies really are "for me."

Such public skepticism toward advanced telecommunications services has been reflected in public opinion polls. For example, the report by Bryant, Gonzenbach, and McCord (1994), which was discussed regarding press coverage of information technology issues, also reported the results of two national public opinion polls about information technology, each featuring a probability sample of approximately 500 adults. The first was conducted in March 1993 and the second in October 1993. Some of the questions on the surveys asked respondents whether they perceived themselves as using various interactive technologies in the future. Figure 6.2 displays the proportion of respondents in each poll who said they expected to use video phones, interactive movies, and interactive television "always," "sometimes," "rarely," or "never" in the future. As can be seen from examining this figure, although approximately 50% of the respondents reportedly saw themselves using video phone "always" or "sometimes" in the future, fewer than 40% reportedly saw themselves as being regular users of interactive

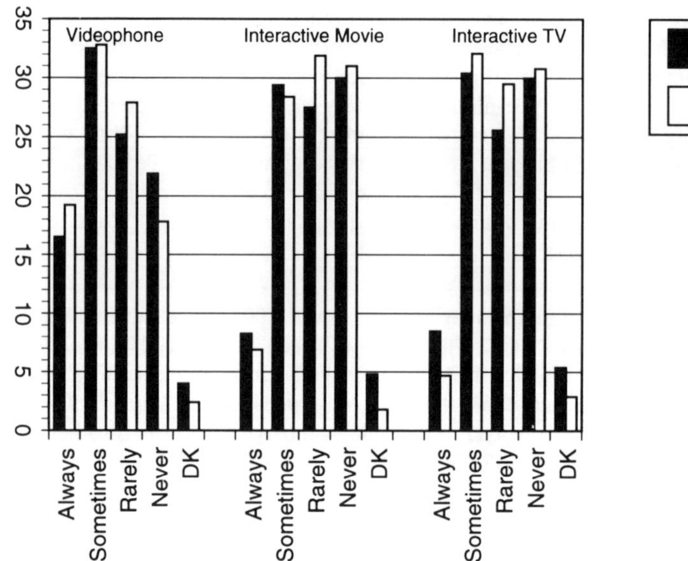

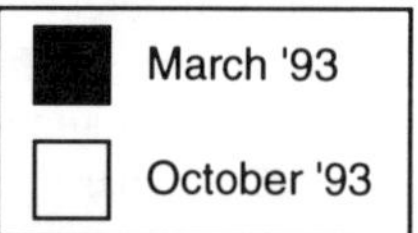

FIG. 6.2. Expected use of videophones and interactive movies and television.

movies or television. Moreover, the passage of half a year between phases 1 and 2 of the survey actually saw optimism about using the latter two services diminish.

Another survey conducted for MCI Telecommunications reported that many of the "glitzy" aspects of advanced information services do not appeal to a large portion of Americans (Gussow, 1994). Of 800 respondents polled nationwide, only 24% reportedly found great appeal in the 500-channel capability of the superhighway, and 35% reported that such information and entertainment abundance had "very little" appeal. Moreover, only 39% found great appeal in the idea of video-on-demand or of interactive television.

A nationwide poll of 1,000 respondents by Clinton Research Services further supports this profile of only modest consumer interest in advanced information services. Only 18% were extremely or very interested in interactive media, and 38% professed no interest whatsoever in interactive services. The conclusion from this survey was that "the gap between marketers' interest and consumer interest is a chasm right now" (Heilbrunn & Branigan, 1994, p. 10).

This apparent lack of consumer interest and optimism, coupled with the major problem of a dearth of killer applications to date, appears to have created a "we'll wait and see but I doubt that it's for me" attitude in the minds of consumers. Such lack of confidence and interest can be a potent barrier to the adoption and use of advanced information technologies in the home.

WHAT PLACE ENTERTAINMENT?

> I find it particularly intriguing that no one has really tried to understand the differences between information consumption and entertainment consumption. The consumers of each are, like you and me, often the same people, yet our motivations for each are different, and our media needs are not the same. (Dickinson, 1994, p. 63)

"Americans spent about $340 billion on entertainment and recreation in 1993. That compares with $270 billion in spending—public and private—on elementary and secondary education. In 1980, those figures were roughly equal" (Landler, 1994, p. 66). Table 6.1 presents the U.S. Department of Commerce data on consumer spending on recreation and entertainment in 1993. Moreover, the rate of spending on entertainment is increasing faster than spending in most other areas; and it is rising faster than the increase in rate of personal income. According to the U.S. Department of Commerce, in 1983 the proportion of consumer spending on entertainment and recreation as a percent of all nonmedical consumer spending was 7.75%; in 1993, it was 9.43%. Or consider that "since 1991, consumers have boosted their outlays on entertainment and recreation by some 13%, adjusted for

TABLE 6.1
Consumer Spending on Recreation and Entertainment (1993) According to the U.S. Department of Commerce

Source	*$(Billions)*
Toys and sporting equipment	$65
TVs, VCRs, recorded music, videotapes, and other media electronics	$58
Books, magazines, and newspapers	$47
Gambling	$28
Cable TV	$19
Amusement parks and other participatory entertainment	$14
Movie admissions and video rentals	$13
Home computers for personal nonbusiness use	$8
Personal pleasure boats and aircraft	$7
Live entertainment other than sports	$6
Spectator sports	$6
Other recreation and entertainment	$70
Total	$341

inflation—more than twice the growth rate of overall consumer spending" (Mandel et al., 1994, p. 59). "One more sign: The rise in entertainment spending has coincided with a dramatic decline in personal savings. In 1980, Americans saved an average of 7.9% of their disposable personal income. In 1993, they saved just 4%" (Landler, 1994, p. 66). Judging from this potpourri of statistics in consumer spending, it would appear that entertainment should well be the perfect driver for new information technology.

Whatever insights we can glean from the development of mass media would seem to teach a similar lesson. The growth of almost all media systems was precipitated by entertainment fare, not by news or information. Book sales grew most rapidly when "dime novels" were introduced. The story of the rise of the mass circulation newspapers is that of the introduction of "yellow journalism," "jazz journalism," and the like. Radio grew rapidly once soap operas, suspense, mystery, and popular music became staple fare. Television cut its teeth on entertainment and obviously remains dependent on such today. The sudden growth rate of VCRs in the early 1980s frequently has been linked to the increased popularity of video pornography, although today the success story of VCRs is that of a shift in the locus of entertainment motion picture consumption from the theater to the home. And how often are compact disc systems, home stereo units, and the like used for news and information? If mass and personal media offer a valid precedent for the place of entertainment in new information technology growth, the case for entertainment as a driver is clearly strengthened (Brown & Bryant, 1989; Bryant, 1989).

Many industry prophets clearly have gotten that message and have issued the call that entertainment must lead the way if advanced information technology is to become a cultural universal in homes, like television and radio, for example, are today. For example, Edward McCracken, CEO of Silicon Graphics, has said that "the entertainment industry is now the driving force for new technology, as defense used to be" (Mandel et al., 1994, p. 60).

Others argue that a blending of entertainment with other types of information will be essential for the economic success of information technology, and they often claim that other forms of information must adopt entertainment embellishments if they are to thrive in the new media environment. Consider education: "But like everything else, education will be different out on the electronic road, and like everything else out there—news, shopping, paying bills—education will be melded with entertainment" (Maurstad, 1994, p. 1C).

Or take advertising: "The feeling and tone of advertising will change from a distant, status-driven relationship to an intimate, playful adventure. In many cases, advertising and entertainment will merge to become 'infotainment' " (MacEvoy, 1994, p. 42).

Or games: "Look for two-way TV to really get serious in the game arena. After all, the appeal of most game shows isn't watching the contestants, but playing along" (Mannes, 1994, p. 46).

Not everyone would agree that advanced information technology should rely on entertainment to be its primary driver. Many have argued forcefully that there is a lack of fit between (a) what most people seem to want from entertainment and how they typically approach entertainment, and (b) the style of media use required by most information technology. They argue for a serious lack of goodness of fit between the less active, disengaged style in which mainstream entertainment has traditionally been served up to audiences (i.e., television, radio, motion pictures) and the more active, selective, interactive, participatory nature of advanced information technology, and they see these differences in style between traditional entertainment consumption and new media use as being a major problem. In other words, they agree with Ted Turner: "Most people want to sit back and watch – interacting is hard work" ("What If They're Right?" 1994, p. 3).

Not according to multimedia evangelists such as Trip Hawkins: "People seem to think we prefer sitting passively in front of television," says Hawkins, president of 3DO Co., which has developed technology for interactive TV. "My opinion is we simply haven't had the alternative. Give viewers the means to interact with the tube, he argues, and you open up new vistas " (Landler, 1994, p. 66).

It may well be that Hawkins is correct, and if so, the general public may not yet be able to accurately judge their future interests in entertainment uses of information technology. Nonetheless, the polls discussed previously are the best indicators we have, and they do to raise legitimate questions about such unbridled optimism. So do the results of a survey of 1,621 Alabama respondents. In recent telephone interviews about factors they consider important in selecting a telecommunications company in a competitive environment (Bryant, Maxwell, Love, Stovall, & Cotter, 1994), respondents were queried about their future decision making. One question asked very specifically, "How important is having additional entertainment or video services available in choosing a telecommunications company?" As can be seen from Figure 6.3, only 14.2% of the respondents considered additional entertainment or video services to be "very important" to them; and 58.1% reportedly thought having such services online was either "not very important" or "not at all important" to their selection of a telecommunications service provider.

This point about apathy regarding additional services was made rather cogently in an episode of "Cheers." Norm, the beer-swilling regular, had his attention captured by a bank of large screen televisions on the wall of a Boston bar. "Well, Normie," said his companion Cliff, "this is the information age. We can get up-to-the-minute stock prices, medical break-

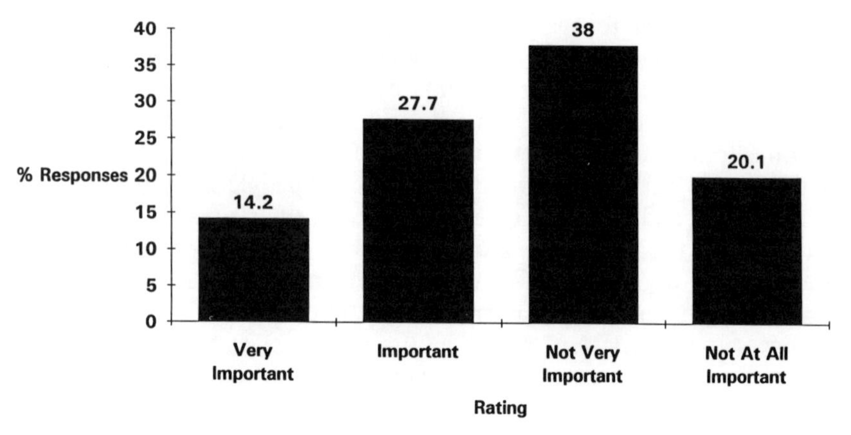

FIG. 6.3. Perceived importance of additional entertainment or video services in choosing a telecommunications company.

throughs, political upheavals from all around the world. Of course, we'd have to turn off the cartoons first" ("What If They're Right?" 1994, p. 3).

Obviously the place of entertainment in marketing of information technology is not an issue about which there is total agreement. Perhaps lessons learned from the myriad industry trials will shed more light on this critical question.

What Place the Home?

> The fastest growing segment of the PC world is that which the industry once wrote off as a wasteland: the home. (Hewson, 1994, p. E12)

Prior to considering those trials, only one more point needs to be made, and that is that the home increasingly is becoming the place where the economic success or failure of information technology adoption and use will be determined. There seems to be widespread industry agreement that the battle has been fought and won on the business front, and the advances to be gained there from now on will be relatively small. The home, in contrast, is the place where the information technology action is, whether the topic be games, or gambling, or education, or video on demand, or advertising, or whatever.

Many commentators note that the ground rules will be very different—and even the playing field will be very different—when the locus of the market becomes the home front. For example:

> For more than a decade, home computers and business machines have walked on different sides of the street: one cheap, underpowered and trapped in a

> variety of incompatible operating systems; the other expensive, hard to learn and lacking in attractive applications that would appeal to a general user. Will this change when more attention is turned to the home market? (Hewson, 1994, p. E12)

> While PC makers had the business market to sustain them during the interim decade, companies in new media for the home may not have the same luxury. "Companies investing billions in this market can't afford to start off on the wrong foot with consumers," said Donatiello. He stressed that the question is not whether this market will develop, but when and how. "Our research clearly indicates that the answer will depend on how effectively companies approach consumers," said Donatiello, "and that will be very different from approaching business customers." ("Nationwide Study Reveals Risks," 1994, p. 1)

> Will the home machine [the access device for the information superhighway] be a "compuvision" or "teleputer?" Clearly we do not yet know, but we must find out if we are going to thrive in the next phase of the new media environment. (Press, 1993, p. 19)

These quotes reflect both certainty and indecision. There appears to be certainty that the home market is becoming more and more crucial for the future of information technology. There appears to be considerable indecision over the shape of things to come in that market. Perhaps that explains why so very many information technology trials are now being conducted in homes. We now turn to an examination of some of those trials and what they can teach us.

INDUSTRY TRIALS

> But the information highway remains shut down as construction work continues. (Rebello, Hof, Weber, & Mallory, 1994, p. 30)

According to some estimates, more than 400 industry trials of new communication technologies and services have been or are being conducted in the United States alone. Obviously it would be impossible to parsimoniously review all of these trials in this volume; moreover, the trade press and several news weeklies are accepting that challenge quite regularly (e.g., Elmer-Dewitt, 1994; Maddox, 1994; Mannes, 1994), and these reports provide much more current information about such tests and pretests than could possibly be provided here. Nonetheless, examining details from a selective sampling of such trials may be instructive, and Table 6.2 presents a profile of a baker's dozen of major information or communication

TABLE 6.2
A Profile of Selected Information Trials and New Entertainment Services

Trial or Product	*Company*	*Application*	*Location*	*Description*
The Edge	Sega of America	Interactive games	Available at retail shops worldwide 1994	Released in the summer of 1994, The Edge (made by AT&T) connects to Sega Genesis players and allows users to play videogames in real time with other players via telephone lines. Estimated cost: $150
EON	EON (formerly TV Answer)	Wireless I-TV	Testing: 1994, Reston and Fair Oaks, VA	EON's wireless two-way communications capability allows home television sets to receive and send information and transactions through a nationwide network of cell sites to, from, and through its central processing facility. Prototype units attach to the viewer's television set and are controlled by a hand-held unit. Services provided include entertainment and educational programming, electronic home shopping, and a variety of other transactional applications.
Full Service Network	Time Warner Entertainment	I-TV	Testing: 1994, Orlando, FL	This fiber-optic system provides a variety of video-on-demand services including movies, sports, news, cultural events; financial services; interactive video games; and electronic shopping.
ImagiNation Network	Imagination	Video games and E-mail	Available nationwide 1994	Using special software, an internal modem, keyboard, and their television set, owners of Panasonic's 3DO multiplayer can join personal computer users to play interactive games and converse online. Estimated cost: $9.95 per month and $3.50 per hour line charges.

Interactive Channel	IT Network (subsidiary of AmeriTech)	I-TV	Testing: 1994, Birmingham, MI and Denton, TX	Subscribers can choose from a variety of interactive services such as video shopping, news and information, and communication with local school systems. Estimated cost: $2 per month.
Interactive Network	Interactive Network, Inc.	I-TV	Tentatively available nationwide 1995	The Interactive Network is a subscription service that allows home viewers to interact in real time with televised game shows, sports, entertainment, and educational programs for fun and/or prizes. The system works independently on TV signals and cable wires by transmitting via an FM radio data channel broadcast by a local station. Using a flat laptop device, subscribers can record their responses then connect the device to their telephone line and call an IN central computer. Results are transmitted back to the unit within minutes. Estimated cost: wireless control unit, $199; monthly charges $10–$15.
Interaxx Machine	Interaxx Television Network	I-TV	Testing: 1994, Coral Springs, FL. Tentatively available to 15 million Channel America subscribers in 1995	The Interaxx Machine utilized 486 PC technology integrated with CD-ROM player, digital coder/decoder, modem, and printer. The device attaches to the home cable line. Via a remote control, users can access a variety of entertainment programs, video games, and online shopping services. Quarterly, two CD-ROM disks are mailed to subscribers: one for shopping, the other for video games. Information contained on the disks can be updated continuously through an out-of-band frequency on the cable line. The Interaxx machine prints out receipts and invoices for products and services as well as vouchers that can be exchanged for tickets to entertainment events. It can also play musical compact disks and CD-ROM games.

(*continued*)

TABLE 6.2 (*continued*)

Trial or Product	*Company*	*Application*	*Location*	*Description*
Main Street	GTE Main Street Inc. (GTE Corp.)	I-TV	Testing began 1989. Available in Cerritos and San Diego County, CA; Boston, MA	Interactive system provides subscribers access to about 50 different services via a telephone line and home TV. Combining audio, text, graphics (but no moving video), Main Street's services include news, sports, weather, and financial information; electronic home shopping; games; airline reservations; and bill payment. Cost: $9.95 per month.
Prevue Express	Prevue Networks (United Video Satellite Group)	I-TV (program navigation)	Testing: 1995	Viewers can search for programs by either date, network, or genre by scrolling through an 8-day program grid. Color coded programs listed in the grid are linked to full motion promotion video clips for movies or other pay-per-view events. Estimated cost: $2–$4 per month, Prevue Express cable box required.
The Sega Channel	Sega of America, TCI, Time Warner	Interactive games	Available nationwide 1994	Cable services that provides dozens of games for Sega Genesis machines by incorporating The Edge device (see previously). Also provides previews of new games still in the developmental stage.
Sports Trakker	Trakker Interactive Services (United Video Satellite Group)	Information service	Testing: 1994, Knoxville, TN	Subscribers receive constantly updated local and national sports scores, weather information, fishing reports, on-screen programming guide.

Stargazer	Bell Atlantic	I-TV	Testing: 1994, northern VA; national rollout 1995	Stargazer is one of the most ambitious trials of video on demand to date. Using a decoder box that attaches to the telephone line and television set, the subscriber calls the Stargazer service and keys in a number corresponding to the requested program. Stargazer uses digital versions of movies for an exceptionally clear picture. Bell Atlantic plans to make the service available to over 1.2 million people by 1995 and over 8 million by the end of the decade.
StarSight Telecast	Viacom Cable and AT&T	I-TV	Testing: 1994; Castro Valley, CA. Tentatively available nationwide by 1995	Using a StarSight remote control device, subscribers can scan a program grid displaying by channel descriptions of television programs airing up to a week in advance. StarSight users can also search for current and future shows by category (i.e., "sports") and subcategory ("i.e., "football"). StarSight also offers one-button VCR recording. The service is accessed by a special decoding device or a TV, VCR, or cable box equipped with the appropriate circuitry. Estimated cost: $4 per month.

technology trials or new entertainment services. In fact, the text of this table reports all that we say about the descriptive details of these trials.

The more important question for a volume like this is, what can we learn from these many exercises in inductive reasoning? Some reporters who have examined the trials closely have been rather skeptical in their evaluations. For example, a *Time* report concluded:

> The faith behind these trials is being sorely tested this spring. Most of them are running behind schedule, many don't work at all, and none are ready for prime time. Not only is the basic technology snarled, but the road ahead is cluttered with legal and regulatory obstacles. Says Mitch Ratcliffe, editor of the newsletter *Digital Media*: "I think consumers are going to be unimpressed for a decade or more." (Elmer-Dewitt, 1994, p. 44)

A *Business Week* report was equally negative:

> We decided we're not going to get enough out of the test at this point. The big question: Will anyone? The flurry of interactive-TV trials is based on one assumption: There will be a huge appetite for movies on demand, and everything beyond that will be frosting on the bottom line. But there's precious little proof. (Rebello et al., 1994, p. 31)

An *Electronic Media* report on the highly publicized GTE Cerritos project concluded simply: " 'Cerritos has bombed,' " said Michael Noll, dean of the Annenberg School of Communication at the University of Southern California. " 'There is no market need for the service' " (Maddox, 1994, p. IM-2).

Interestingly, all of those reports concluded at least one section of their evaluation with the phrase, "not ready for prime time!," utilizing a mass medium metaphor to pan the results of information technology trials.

Because of the inductive nature of these trials, such sweeping conclusions are not very useful. Perhaps more valuable results can be gleaned from the "nitty-gritty." What is provided is an exhaustive potpourri of quotes from all the results we could gather from reports of these trials:

> What GTE has learned is how to put together the components of a network and understand what customers want. "They like choice, convenience and service," he said.
>
> "We've learned a lot about the nuances of what people like, and not just in terms of liking 30 channels of pay per view or choosing from 1,500 movie titles," said Dede Moreland, marketing manager of VCTV. . . . "We've learned what people like in terms of how simple or complex the ordering screens need to be, how they like the user interfaces, and what types of movie titles they want to buy," said Moreland. . . . "We have learned that we will

need to add a new feature that will give viewers a 15-minute 'intermission' on the expanded pay-per-view service."

The navigational system is the most important feature of the network.

In general, consumers want the services they already have enhanced and made more user friendly.

That makes video on demand the most obvious application, he said, as opposed to existing pay-per-view services, followed by interactive programming guides and whole-house TV services, where all the TVs are hooked up to the network.

The entree to the consumer is to get existing problems fixed, to get it user friendly and to get it to the home. (Maddox, 1994, pp. IM 2-4; 14)

Some trials indicate consumers won't pay more than a $1 premium over a store rental for an on-demand movie. (Fatsis, 1994, p. 8)

With more than 60 video games available on the system, the average household use is five hours a week. (Johns, 1994, p. 25)

The bread-and-butter interactive TV applications are movies and home shopping. (Press, 1994, p. 19)

It's most popular features are games, like blackjack, where subscribers play each other, and the educational services. (Frank, 1994, p. 1)

Initial response from viewers offered "movie-on-demand" services has been mixed, with most consumers making little use of it. (Kehoe, 1994, p. V)

Needless to say, both the quality and quantity of this catalogue of results are quite disappointing. Why is the list so meager? A major portion of the reason undoubtedly is that these are commercial trials, and presumably considerable profit potential is at stake, so the "real" results are proprietary. In fact, what has been reported to date in the trade press, business publications, and news magazines is: (a) a great deal of hype about what so-and-so is planning in terms of a new information technology or service trial; (b) a modicum of material about how the trial is progressing, typically including perfunctory accounts about or vapid interviews with users of the technologies and services; and (c) next to nothing about the results of the trials. Follow-up telephone calls, even to sources listed in press releases as the contact person for the trial, proved to be equally unrevealing. It is obvious that either (a) the results have been inconclusive or disappointing, or (b) management has slapped a proprietary label on the findings.

Interestingly enough, searches for useful reports from university laboratories where some of the research and development for these projects supposedly took place proved to be just as fruitless. Needless to say, what we have learned to date from industry trials is that you can get a lot of press by announcing a trial, but neither media nor scholars seem to be able to glean very much information from the results that would help us understand the place of entertainment in selling advanced information technology to consumers or how consumers are using or benefiting from such technology and services.

SPECIAL ATTRIBUTES OF NEW INFORMATION TECHNOLOGY THAT COULD ENHANCE THE ENTERTAINMENT EXPERIENCE

When we set out to write this chapter, our plan was to balance the evidence from *inductive reasoning* (e.g., the industry trials) with insights garnered from *deductive reasoning* (e.g., theoretical rationales), which should together lead to explanations and predictions of the potential of entertainment as a driver of new media adoption and use. Unfortunately, a search of the extant literature in relevant domains (e.g., entertainment theory, media uses and effects, consumer behavior) revealed a dearth of directly applicable theory. In fact, we decided that "new technology theory" was practically an oxymoron, at least as far as theory leading to fuller understanding of the psychology of consumer behavior was concerned. Therefore, we redefined our task and decided to create some pre-theoretical orienting statements that could help lay the foundation for theory development that could help answer the question we were raising in this area.

One way to begin to generate this type of social scientific theory is to identify key elements of human behavior that are related to the entertainment experience, specifically those that are likely to be altered via the use of advanced information technology. Another way to look at this is to questions what human potential new media unleash that is not present in traditional mediated entertainment (cf. Bryant, 1993). What one should look for in this regard are shifts that represent a real change—often defined by Gregory Bateson as "a difference that makes a difference." Some such changes and a sampling of entertainment theory questions they raise are as follows:

Selectivity: In the often referenced 500-channel environment, *selectivity* is practically forced on the system user. No longer are users limited to the choice between three remarkably similar sitcoms on ABC, CBS, or NBC. In the future, we will be selecting from scores of different options within each of perhaps a dozen differerent genres, some of which may be new or at least hybrids. Psychologically, how will we respond when we can narrow our choices to media we really like, to genres we prefer, to characters we really care about, to settings and situations that are germane to our lives—or to our secret fantasy lives? Will this potential for selectivity broaden or deepen our entertainment experience? Will it alter our media use in significant ways?

Diet: Closely related to selectivity is the notion of *diet.* Diet refers to the composition of information that users select and can include variations in type of programming (e.g., entertainment, education, or information), between types of media (e.g., print vs. electronic), between reception formats (e.g., audio, video, AV), and the like. Given that we have access to a smorgasbord of entertainment fare, what diet plan will we follow? Will we binge on our favorite sweets? Will usage be biased toward light, easily digestible fare? Will entertainment bulimia or anorexia result? Will we quickly become satiated with one type of treat and then graze on more varied, health fare?

Interactivity: The concept of *interactivity* is that element of information technology most often discussed in the trade and popular presses, and logically the capacity to respond to other messages, act on them, and the like should alter the entertainment experience qualitatively. But will we even choose the way of interaction? Will all of us? How often? Under what conditions? To what effects? How will this alter our engagement or involvement? Our attention? Our information processing? Our comprehension?

Agency: *Agency* refers to the degrees of control one possesses in relation to a technology or system. The term can be used as a psychological personality factor (e.g., locus of control) or it can be referred to as the potential of the medium for user empowerment. If an information technology affords such agency to the user that basic aspects of message structure (e.g., the language or accent employed) can be altered, will this alter the entertainment experience? Will we really want to select the denouement or resolutions of our dramas, or to select the personalities and physical characteristics of our soap opera heroes, or to alter the lyrics to our popular music? If so, will this shift the boundaries of that we call "entertainment?"

Personalization: To a certain extent, a combination of agency and interactivity are attributes of *personalization.* Mass media messages typically have been addressed "to whom it may concern." Direct marketing innovations have caused consumers to expect quasipersonalization (e.g., not "Dear Occupant," but "Dear Fred and Wilma Flintstone"). The next generations of media will permit true customization of messages along a number of lines. In entertainment, how much personalization is desired? How salient is too salient? How real is too real? Is personalization the enemy of the willing suspension of disbelief, long believed to be essential to certain types of entertainment?

Dimensionality: In the digital world, messages are not only infinitely malleable, they can have many different dimensions. Looked at overly simply, will we always desire a *multi*media experience, or will we sometimes prefer to only read text, or to only listen to music without the videos? How much multidimensionality is desirable for maximal enjoyment under what context and social situation?

The list certainly could continue, beginning with characteristics like transparency (or user friendliness) or accessibility. And just as we complete the list, some innovator will create a new information technology or service that causes us to expand the list. This is the blessing and the curse of living in the information age.

A PROPOSAL FOR DOING THEORY TRIALS

Another emerging characteristic of living in an information society is the availability of collaboratories. The potential to use information technologies to interconnect scholars and practitioners, technologists and specialists in human behavior, and many other diverse "species" of problem solver leads us to recommend a new model for asking and answering questions like, "What is the place of entertainment as a driver of information technology in the home?"

What is needed is definitely not another atheoretical trial stemming from a simplistic marketing response to an engineering innovation. What is needed first is a productive collaboration at the theoretical level—a creative discussion of entertainment theory in union with insightful analyses from diverse sources of which dimensions of new information technologies are likely to contribute to an enhanced entertainment experience. In other words, first we need to create something as practical as good entertainment theory for the information age. Then our theories can generate good research questions that can lead to logical, substantive trials that are not only market-sensitive but consumer-driven in the best possible sense, with the goal of maximizing the potential of new information technologies to enhance "the pleasure principle." This fusion of deductive and inductive reasoning potentially will provide a much more valid and durable approach to, first, asking the right questions and then, turning our research expertise and technological resources toward providing the most useful solutions to these very important questions. Other approaches capitalize only on a small portion of the many ways we have learned how to learn.

REFERENCES

Andrews, E. L. (1994, January 3). Big risk and cost seen in creating data superhighway. *The New York Times*, p. 17.

Andrews, P. (1994, February 8). More realistic view of "information superhighway" is taking shape. *The Seattle Times*, p. D2.

Brown, D., & Bryant, J. (1989). An annotated statistical abstract of communication media in the United States. In J. S. Salvaggio & J. Bryant (Eds.), *Media use in the information age: Emerging patterns of adoption and consumer use* (pp. 259-302). Hillsdale, NJ: Lawrence Erlbaum Associates.

Bryant, J. (1989). Message features and entertainment effects. In J. J. Bradac (Ed.), *Messages in communication science: Contemporary approaches to the study of effects* (pp. 231-262). Newbury Park, CA: Sage.

Bryant, J. (1993). Will traditional media research paradigms be obsolete in the era of intelligent communication networks? In P. Gaunt (Ed.), *Beyond agendas: New directions in communication research* (pp. 149-167). Westport, CT: Greenwood Press.

Bryant, J., Gonzenbach, W. J., & McCord, L. (1994, May). *Press coverage of and public attitudes toward the information highway*. Paper presented at a conference of the World Association of Public Opinion Research, Danvers, MA.

Bryant, J., Maxwell, M., Love, C., Stovall, J. G., & Cotter, P. R. (1994). *South Central Bell benchmark survey: April 24-May 21, 1994*. Tuscaloosa, AL: Institute for Communication Research. [Proprietary Report. Selected results included with the permission of South Central Bell.]

Bureau of National Affairs. (1994, January 12). *Daily Report for Executives*, p. N8.

Coates, J. (1994, February 13). If you can't beat 'em, modem: The computer revolution has hit home. But is everyone at home worth it? *Chicago Tribune*, p. C10.

Dickinson, J. (1994, January). What will convergence bring us? Merging of computer and television technology requires more thought. *Computer Shopper*, p. 63.

Downs, A. (1972). Up and down with ecology: The "issue-attention cycle." *Public Interest, 28*, 28-50.

Elmer-Dewitt, P. (1994, May 23). Play . . . fast forward . . . rewind . . . pause. *Time*, pp. 44-45.

Fatsis, S. (1994, March 1). Hype tends to obscure the meaning of interactive TV. *Los Angeles Times*, p. 8.

Fishman, G. (1994, February 5). Slow motion on info superhighway. *Billboard*, p. 4.

Frank, R. (1994, March 16). Television, the total services pipeline. *International Herald Tribune*, p. 1.

Frankel, B. (1993, December 13). Coming to a set near you. *Adweek* (Eastern Edition), p. 28.

From prime time to my time. (1994, February 12). *The Economist*, p. 9.

Gussow, D. (1994, January 17). "Interactive" TV means "active" to U.S., poll says. *St. Petersburg Times*, p. 10.

Hamilton, J. H. (1994, April 10). Rate of change grows ever faster. *The San Francisco Examiner*, p. C5.

Heilbrunn, H., & Branigan, L. J. (1994, February 21). Consumers just don't get it; marketers who plan to use interactive media to sell to a public of technology tenderfoots have a lot of explaining to do. *Mediaweek*, p. 10.

Hewson, D. (1994, February 20). Battle shifts to the home front. *Sunday Times*, p. E12.

It's the end of the world as we know it. (1994, February 12). *The Economist*, p. 17.

Johns, A. (1994, February). Hearst goes interactive, launches services for the home. *Digital Media*, p. 25.

Kehoe, L. (1994, March 8). Survey of US communications. *Financial Times*, p. 5.

Landler, M. (1994, March 14). Are we having fun yet? Maybe too much. *Business Week,* p. 66.

Lohr, S. (1994, March 1). The silver disk may soon eclipse the silver screen. *The New York Times,* p. A1.

MacEvoy, B. (1994). Change leader and the new media. *American Demographics,* p. 42.

Maddox, K. (1994, March 21). Setbacks on the superhighway: A status report on who's testing what in 1994, the year of interactivity. *Electronic Media,* IM-2-4; 14.

Mandel, M. J. (1994). The digital juggernaut. *Business Week Special Issue: The Information Revolution,* pp. 22–31.

Mandel, M. J., Landler, M., Grover, R., DeGeorge, G., Weber, J., & Rebello, K. (1994, March 14). The entertainment economy: America's growth engines: Theme parks, casinos, sports, interactive TV. *Business Week,* pp. 58–64.

Mannes, G. (1994, January). Two-way TV, the year's best. *Video Magazine,* p. 46.

Maurstad, T. (1994, January 16). On the superhighway, the rides are sheer joy. *The Dallas Morning News,* p. 1C.

Nationwide study reveals risks of current approach to marketing new media.(1994, February 17). *Business Wire,* p. 1.

Neuman, W. R. (1990). The threshold of public attention. *Public Opinion Quarterly, 54*(2), 159–176.

Press, L. (1993, December). The Internet and interactive television. *Communications of the ACM,* p. 19.

Rebello, K., Hof, R. D., Weber, J., & Mallory, M. (1994, March 14). Interactive TV: Not ready for prime time. *Business Week,* pp. 30–31.

Salvaggio, J. S., & Bryant, J. (Eds.). (1989). *Media use in the information age: Emerging patterns of adoption and consumer use.* Hillsdale, NJ: Lawrence Erlbaum Associates.

Stern, A. (1994, February 8). The shape of advertising to come on the information highway. *The Reuters Business Report,* p. 4.

Technology. (1994, January 28). *Daily Report for Executives,* NS18.

What if they're right? (1994, February 12). *The Economist,* p. 3.

CHAPTER 7
An Approach to Mapping Entertainment Alternatives

Gregory W. Cermak
GTE Laboratories Incorporated

The Market Context for New Information Technologies and Product Maps

Consumer acceptance of new information technologies heavily depends on the market context in which the technologies are introduced (e.g., Jain, 1993). An important element of that market context is the consuming public's framing of the competitive set: Which current *products*[1] do consumers view as sources of information, and under what circumstances? For example, when screen phones[2] are introduced, some consumers may view them as falling in the same competitive set as high-end wall phones and speaker phones, whereas other consumers may place them in a competitive set containing personal computers and personal digital assistants.

The market context includes the attributes by which consumers judge current information products (also see Jain, 1993). Consumers will most likely think of new technologies in terms of familiar attributes; those attributes describe and differentiate existing technologies. For example, the attributes that characterize residential telephone equipment (e.g., style, price, number of buttons) can still be used by consumers to understand the new screen phone entrant. The attributes for one competitive set may differ from the attributes of another competitive set. For example, if consumers perceive the competitive set for screen phones to include personal computers and personal digital assistants, then attributes such as price, footprint, portability, screen size, and memory might become important.

Many tools are available for describing the market context of products based on new technologies in terms of both product set and comparable attributes. Making sense of data concerning the competitive relationships

[1]In this chapter *product* is used in its broad form to refer to both goods and services for sale.

[2]Screen phones have visual displays for text and possibly simple graphics, but not video, and programmable keys or a touch screen. They often include a pull-out keyboard.

among many existing products, however, can be difficult. Fortunately, making sense of relationship data is precisely the strength of a product map (Green, 1975; Shocker & Srinivasan, 1979; Stefflre, 1986).[3] A product map, like a road map, is a tool for the decision maker. As Shepard (1972a), the originator of the data analysis technique for product maps, multidimensional scaling (MDS), put it: "The unifying purpose that these [multidimensional scaling] techniques share, despite their diversity, is the double one (a) of somehow getting hold of whatever pattern or structure may otherwise lie hidden in a matrix of empirical data and (b) of representing that structure in a form that is much more accessible to the human eye—namely, as a geometrical model or picture" (p. 1). A product map is thus a useful tool for understanding the current product context and thereby anticipating demand for new technologies.

Note that a product map is a useful tool in the same sense that other maps are tools: Product maps do not function by themselves. Product maps do not provide the product manager with definitive plans, but rather provide the background information on which plans can be based. By analogy, a map of enemy positions does not, by itself, recommend military strategy and tactics. But, as information in the hands of someone commanding resources, maps are indispensible for stimulating plans and informing decisions. As the commander of military forces is informed by a map of the terrain and the disposition of enemy forces, so is the product manager informed of the market context and location of competitors by a product map.

Information and Entertainment

Although information and entertainment may be conceptually distinct, from a practical point of view they are hopelessly tangled. One would be hard pressed to name a vehicle for delivering information that is not also used for delivering entertainment. Print and electronic media, both video and audio, serve both functions, very often simultaneously. In the screen phone example, since residential telephones and personal computers are used for both entertainment and information, one would predict that screen phones would also be used both ways as well. One might argue that the concept of entertainment is broader than the concept of information in that humans can derive enjoyment from activities that have little obvious information content: playing softball or dancing. Alternatively, one can view "information" as the broader category as in cases such as seeing

[3]Product maps are multidimensional figures that locate an array of products according to either (a) judged ratings on key attributes, or (b) judged similarities or substitutabilities among the products.

crossing signals at an intersection or hearing the directions to a retail store. However, often entertainment becomes a consequence of information transmission as in reading a book or listening to music where information and entertainment are inseparable. This chapter assumes this broader point of view, although the study and maps reported here were originally conceived in an entertainment context.

Previous Maps

Effective product maps, of information sources or entertainment sources, must be based on data. The data that are easiest to obtain are not always the data one wants. In the case of entertainment products, maps do exist, but, as maps of product competition, they are all inappropriate in one way or another. The existing map that comes closest to providing a tool for the entertainment product planner was developed explicitly to represent complementary relationships rather than substitutable relationships (Holbrook & Lehmann, 1981). In a complementarity map, proximity means that two products tend to be used by the same consumer, possibly jointly. For example, in the case of information products, the same consumer may (a) read the *Wall Street Journal* (*WSJ*) and listen to the marine weather report on the radio, or (b) read the evening TV listings and listen to the deer hunting report on the local am radio station. The *WSJ* and marine weather reports are not competitors, nor are they similar in most of their attributes, save that they might appeal to the same consumer. Similarly, the TV listings and radio hunting reports share little except their audience. (For further discussion of previous entertainment maps, see the appendix.)

However, data for such maps can be collected efficiently for large numbers of products. In the case of entertainment sources, maps are often based on a consumer-by-product usage inventory (Bishop, 1970; Duncan, 1978; Holbrook & Lehmann, 1981; McKechnie, 1974). A respondent indicates amount of usage for each of perhaps 100 products or activities on a checklist. In this approach, products or activities are related by virtue of being used by the same person. In fact, the entertainment studies were traditionally meant to address consumer issues, not product issues. The data emphasize within-respondent similarities and between-respondent differences, as opposed to product similarities. The resulting maps reflect dimensions that emphasize individual differences in style or gender. Maps of information products built on consumer-by-usage data would also be prone to emphasizing consumer similarities and differences, rather than product similarities and differences.

Another unfortunate characteristic of typical usage data for entertainment activities is that the activities or products included need not have been used at the same time. This characteristic de-emphasizes situational similarity of product usage and also weakens the interpretation of complementarity. Strong arguments have been made that product competition depends heavily on substitution in use, and hence on similarity of usage situation (Day, Shocker, & Srivastava, 1979; Ratneshwar & Shocker, 1991).

Another class of maps does use data that might be interpreted as similarity or substitutability, but unfortunately the individual maps contain too few products and activities to be useful as a broad market overview (Becker, 1976; Hirschman, 1985; London, Crandall, & Fitzgibbons, 1977; Ritchie, 1975). The data on which such maps are based are usually judgments of similarity for all pairs of a limited set of products. Since the number of pairs increases as roughly the square of the number of items, such studies rarely include more than 15 or 20 products. This can be a significant problem for studying the market context for new products based on new technologies because the new products may have attributes that are quite different in degree or kind from the attributes of a small set of existing products.

The study presented in this chapter helps the product planner understand the market context for information products, and especially the overlap of information products and entertainment sources. The maps presented show which current information and entertainment products are competitors in the eyes of consumers. The maps also indicate the high-level attributes that differentiate among information products and among entertainment products.

The maps' information on competition and high-level attributes will help form a basis for decisions in three areas: (a) development of new information products, (b) choice among new product concepts, and (c) positioning. Understanding possible competitors and high-level attributes will assist the product planner in determining how new information products can be differentiated from existing products in both the information and entertainment domains. Information and entertainment naturally overlap considerably, thus the maps should also suggest ways in which this overlap can be used to advantage, rather than coming as an unpleasant surprise. Finally, the methodology presented here is adaptable: It can be used specifically for information products in future studies.

METHOD

Generating Lists of Information Sources

To generate a list of information sources, or, in the case of the present study, entertainment sources, we modified an interviewing technique known as *laddering* (Reynolds & Gutman, 1988). In one-on-one interviews, a consumer first named common entertainment sources. Then, the consumer compared pairs of sources with respect to their attributes and benefits. Using the attributes and benefits, consumers then named additional sources that provide similar benefits or that have similar attributes. In this way, we acquired an expanded list of entertainment sources that were united, in the views of consumers, by sharing attributes and benefits.

Approach to Substitutability

Judgmental Versus Behavioral Substitutability. The basis for a product map is data on the relationship of one product to another. A fundamental relationship for products is that of substitutability, the extent to which one product could be used in place of another in similar circumstances. It is this relationship that any new information technology or product will assume with respect to the existing market context when it is introduced. Various sorts of data are more or less perfect indicators of suitability.[4]

Consumers' actions in the marketplace provide substitutability data with good face validity. However, observing actions in the marketplace is time-consuming and expensive, and the number and range of products included in any single study is likely to be quite small—suitable for detailed study of competitive relationships within a narrow product category, but not suitable for generating a comprehensive view of the product domain (e.g., Allenby, 1989; Lehmann, 1972). Of course, even with marketplace data, one normally does not *directly* observe a substitution, but infers it from interpurchase times (Fraser & Bradford, 1983) or interpurchase transition frequency (Grover & Dillon, 1985).

Probably the strongest summary measure of substitutability is cross-elasticity of demand (Allenby, 1989; Day et al., 1979; Holbrook & Lehmann, 1981). In the case of frequently purchased packaged goods, diary or scanner data can document changes in purchase patterns following a price change for a given brand (see, e.g., Krishnamurthi & Raj, 1991; Walters, 1991). In the case of information sources, or of entertainment products and activities, scanner data are not available. In addition, both scanner studies and diary studies are practical only for small numbers of products.

An alternative to marketplace data for indicating substitutability is the class of judgmental measures (Bucklin & Srinivasan, 1991; Day et al., 1979; Holbrook & Lehmann, 1981; Simonson, 1990). As Fraser and Bradford put it (1983): "To infer the competitive structure of a market, one can simply

[4]Previous entertainment maps have not been based on this relationship (see section on previous maps and the appendix).

ask consumers which items they perceive to be substitutes" (p. 27). Among the many types of judgmental measures, one might distinguish (a) direct from derived (or "profile") measures and (b) proximity data from dominance data (Shepard, 1972b). The elements of a profile could be either product characteristics or usages and usage contexts (e.g., Srivastava, Alpert, & Shocker, 1984). A derived measure of substitutability for two products is some function of the products' profiles, for example a correlation or some sort of distance. This function essentially constitutes a theory of substitutability: The elements in the profile are the implied precursors of substitutability, and their relative contribution to overall substitutability is given by the function.

Direct judgments of substitutability, by contrast, do not require the experimenter to assume a theory of substitution. The present study had an empirical aim, rather than a theoretical one, namely to produce a map of entertainment sources. The study employs a direct measure of substitutability rather than a derived one in order to avoid the issue of a theory of substitutability.

Substitutability, similarity, and confusion are all examples of a general proximity relation among items. Choice, preference, and frequency of use are examples of a second kind of relation, dominance (see Shepard, 1972b). Proximity data and dominance data (as well as profile data) can both be used to create product maps (e.g., Carroll, 1972; DeSarbo & Hoffman, 1987). The two types of data are themselves somewhat substitutable in the sense that dominance data can be used to infer proximity relations among items or products, given additional assumptions, and proximity data can be used to infer dominance relations, given some additional information, such as an "ideal point" (Carroll, 1972). Should one use proximity data or dominance data for a product map? One can claim that dominance data occupy a privileged position in social science (e.g., Carroll & De Soete, 1991); one can claim the same for proximity data (e.g., Shepard, 1987). Since what the eye sees in a product map are proximity relations, the present study opts for the directness of using proximity data to construct a map of proximity relations.

Judgment Data Via Sorting. The product manager may be uncertain about which current products will compete with a novel product; a product map with broad coverage (many products) is a hedge against this uncertainty. For example, the screen phone may compete with standard residential telephone equipment, or with office equipment, or with personal computers. Services offered over a screen phone might compete with retail stores, branch banks, broadcast TV programs, or paper books. Given this uncertainty, the product manager would benefit from a map with broad enough coverage to apply in all cases. Several different data collection procedures yield proximity data (e.g., Arabie & Soli, 1982), but one that is especially well suited to large numbers of products is sorting.

The first references to sorting for proximity data seem to be Miller (1967); Stringer (1967); and Rosenberg, Nelson, and Vivekananthan (1968) in psychology, and Green, Carmone, and Fox (1969) in market research. Thereafter, sorting received only occasional attention (e.g., Arabie & Soli, 1982; DeSarbo, Jedidi, & Johnson, 1991; Rao & Katz, 1971; Rosenberg, 1982). Major reviews of MDS, and of MDS specifically in market research, mention sorting as a data-collection method only in passing or not at all (see Carroll & Arabie, 1980; Cooper, 1983; Green, 1975; Punj & Stewart, 1983; Shepard, 1972b).

Compared to proximity judgments of all pairs of products, or to multiple attribute ratings for each product, sorting provides low-density information per respondent. But, each respondent can judge a large number of products. Any single respondent's data contains too little information to constrain an MDS solution (see Arabie & Soli, 1982), but when many data sets are aggregated one obtains data with enough structure to permit an MDS analysis. For large product sets, where respondent fatigue can play a significant role, sorting provides a workable alternative to methods that require more of each respondent (DeSarbo et al., 1991).

Because multiple respondents' data must be aggregated, the resulting product map is best interpreted as a map of the market, not as a map of a mind. That is, the data do not represent any single person's judgments. The data may provide a fair representation of a central tendency of people's judgments, if judgments were distributed unimodally in some multivariate space. On the other hand, if people's judgments were distributed multimodally, the data could provide a poor representation of the various individuals (Anderson, 1973; Day, Deutscher, & Ryans, 1976; Lehmann, 1972). That is, the notion of a map of an aggregate mind might not make sense.

Irrespective of whether or not aggregate data represent any individual, one can still view the aggregate data, and a map based on them, as representing potential substitutability among a set of products. If two products are sorted together by 90% or 10% of the respondents, there is no difficulty in interpreting the aggregate data. If two products are sorted together by 50% of the respondents, one can view that as indicating a moderate potential to compete across a population of consumers, whether that potential is realized as strong competition for half the consumers or as moderate competition for all the consumers. Such a coarse view of the market may not be appropriate for, say, forecasting, but can be quite adequate for getting an initial view of the overall structure of a product domain.

Sample

Forty respondents provided complete data sets. Twenty respondents were nonstudents, equally divided between men and women. Median reported age was 35, and median reported annual family income was $45,000. Twenty respondents were students at a business college, again equally divided between men and women. The respondent sample thus allowed for tests of gross demographic effects of gender and the demographic differences implicit in the student versus nonstudent contrast. Another 25 nonstudents of roughly the same age and income participated in the preliminary one-on-one interviews. The respondents resided in middle-class suburbs of a northeastern city.

Products and Activities Sorted

The one-on-one interviews with 25 respondents identified 149 products and activities that respondents used or engaged in for entertainment. This list of entertainment sources was broad enough to include many information products and activities, as well as activities that also are associated with work (e.g., cooking, yard care), but which might still compete with commercial entertainment products for consumers' time and budget. The list of entertainment sources is least representative of products and activities that respondents hesitate to discuss in public (e.g., activities that are illegal, immoral, or private). Each entertainment or information source was described as a word or short phrase printed on a card.

Sorting Task

The definition of substitutability used here was similar in spirit to the substitution-in-use definition by Shocker and colleagues (e.g., Day et al., 1979; Ratneshwar & Shocker, 1991). Respondents first considered using a given entertainment source, which established an implicit usage context and usage situation (e.g., "sewing" implies a context of at home and probably alone; "attending concerts" implies a context of outside the home, probably at night on the weekend, and probably with others). Then, the respondent was asked to imagine that the entertainment source became unavailable. In that case, what would the respondent do instead? (see Simonson, 1990, for a similar definition of substitutability).

Note that this operational definition of substitutability says nothing about the causes of substitutability, such as the particular benefits or uses of an entertainment source. The definition also is silent concerning the relationship between similarity and substitutability. Finally, the definition avoids the sort of complementarity that a respondents-by-products matrix emphasizes, such as reading the *Wall Street Journal* and sailing, or hunting and watching televised sports.

Respondents participated either singly or in groups of up to three. Each respondent had a deck of 149 cards, each of which carried the name of one entertainment source. Respondents were asked to find one entertainment they enjoyed frequently and place that card on the table. For each succeeding card in the deck, a respondent decided whether the entertainment listed on the card on the table would be a good substitute for the entertainment listed on the card in hand.[5] If the entertainment listed on the card on the table was a good substitute, in the sense defined previously, then the respondent added the card in hand to the card on the table.[6] Otherwise, the card in hand was placed in a new stack on the table. The respondent iterated through this procedure until all cards had been placed in stacks on the table. The number of stacks was left to the respondent. The procedure took about 45 minutes per respondent.

Analysis Procedures

Each respondent's data consisted of 149 assignment of products to stacks, where the number of stacks could differ from respondent to respondent. These data were transformed into a 149 × 149 co-occurrence matrix, coded so that 0 was entered if two products fell in the same stack and 1 was entered otherwise. Note that any respondent's data fit this format, no matter how many stacks they used or what the stacks' compositions were. The respondents' data were aggregated by simply summing the individual matrices (see Carroll & Arabie, 1983; Drasgow & Jones, 1979). The result was a square matrix of distancelike entries (Miller, 1969) with 149 rows and columns, and with values in individual cells that could range from 0 to 40 (the number of respondents). Product pairs with smaller data values were sorted together by more respondents, while pairs with larger data values were sorted together by fewer respondents.[7]

The primary analysis was via SYSTAT's nonmetric MDS routine (version 5.0; Wilkinson, 1991). Following the advice of Arabie, Carroll, and

---

[5]If the card in hand listed a form of entertainment that the subject did not enjoy, the subject was instructed to think of a relative or public figure who might enjoy that entertainment, and think of what that person might consider a suitable substitute. Observation of the subjects and discussion with them indicated that they were able to make reasonable judgments for forms of entertainment in which they themselves did not engage.

[6]Subjects were told that a card could be placed in a stack if any member of the stack were a good substitute for the card in hand.

[7]These data are reliable and consistent across subject groups: Similar but heterogeneous groups of 20 subjects each produced proximity matrices that correlated (product-moment) 0.85; a statistic designed for proximity data (Schultz & Hubert, 1976) was highly significant. The data matrix for 20 male subjects correlated 0.84 with the data matrix for 20 female subjects, and the data for 20 college students correlated 0.81 with the data for nonstudents. The Schultz and Hubert (1976) statistic was highly significant in all cases.

DeSarbo (1987) a cluster analysis was also included. The present analysis used Ward's method for hierarchical clustering (see Punj & Stewart, 1983) in SAS Proc Cluster (SAS, 1985).[8]

RESULTS

Clustering

The market context for information and entertainment sources is very broad. The set of 149 products and activities uncovered in this study is at once difficult to comprehend, yet still not as large as it might be. For each product or activity, one could name another not named here.[9] The clustering aims at consolidating the data for easier interpretation. Statistically, a 14-cluster solution was chosen on the basis of relative decrease in the semipartial *R*-squared reported in SAS' Proc Cluster. The 14-cluster solution accounted for 57% of the variance in the original data. A plot of the hierarchical tree from the clustering is shown in Fig. 7.1

A main split in the tree was between forms of exercise (outdoor sports, exercise and sports) and all other clusters. Within the main body of clusters, other major splits were (a) electronic entertainment sources versus all else; (b) singles entertainment (live events, socializing) versus all else; (c) family entertainment (the nine clusters at the top in Fig. 7.1); (d) within-family entertainment, a split between entertainment without practical output versus hobbies (i.e., reading, social clubs vs. crafts, cars); and, within each of these groups, splits between indoor and outdoor activities (e.g., household, crafts vs. nature, cars). This clustering provides a first-order interpretation of the large quantity of data. The clustering also helps increase the comprehensibility of the MDS product map (Fig. 7.2): The 149 individual entertainment sources are coded so that members of the same cluster are represented by the same symbol.[10]

MDS

The MDS representation of the data was obtained using the parameters: Kruskal's stress form 1, monotone regression, and Euclidean distances.

[8]More powerful clustering methods exist (e.g., DeSarbo, Jedidi, & Johnson, 1991), but, as Arabie and Hubert (1992) noted, "Such developments, however, are not buttressed by the software found in statistical packages, and the result is a widening gap between elegant developments in algorithms and models versus access to them by potential users" (p. 189).

[9]The procedure used for eliciting entertainment sources was designed so that the *range* of named sources would be as broad as possible, although not every source within that range might be named.

[10]The names and coordinates of the individual entertainment sources are available from the author.

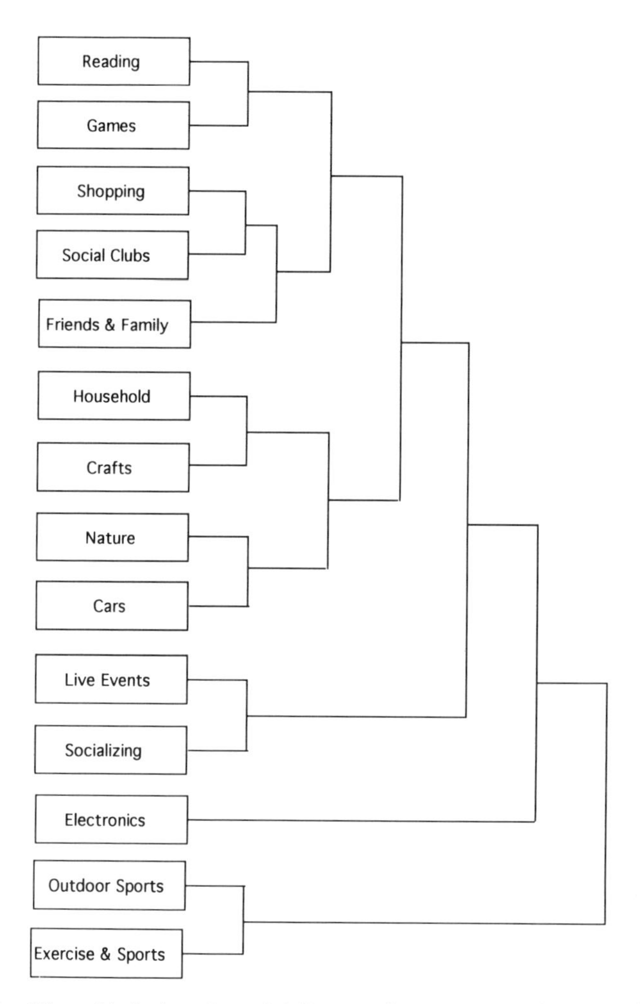

FIG. 7.1. Hierarchical clustering of 149 entertainment sources based on judged substitutability.

Solutions were computed in two through five dimensions. By reference to plots of expected stress values (De Leeuw & Stoop, 1984), the data were nonrandom, but the choice of proper dimensionality of the solution was not obvious. The two-dimensional solution (stress = 0.203 for 149 points) is presented in Fig. 7.2. Clustering information is included in the map by assigning the same symbol to each product in a cluster. Note that the clusters are reasonably compact in the map.

As in the tree (Fig. 7.1), exercise and sports are distant from the center of the map, as are the electronic forms of entertainment. The map captures the additional information that exercise and electronic entertainments are not

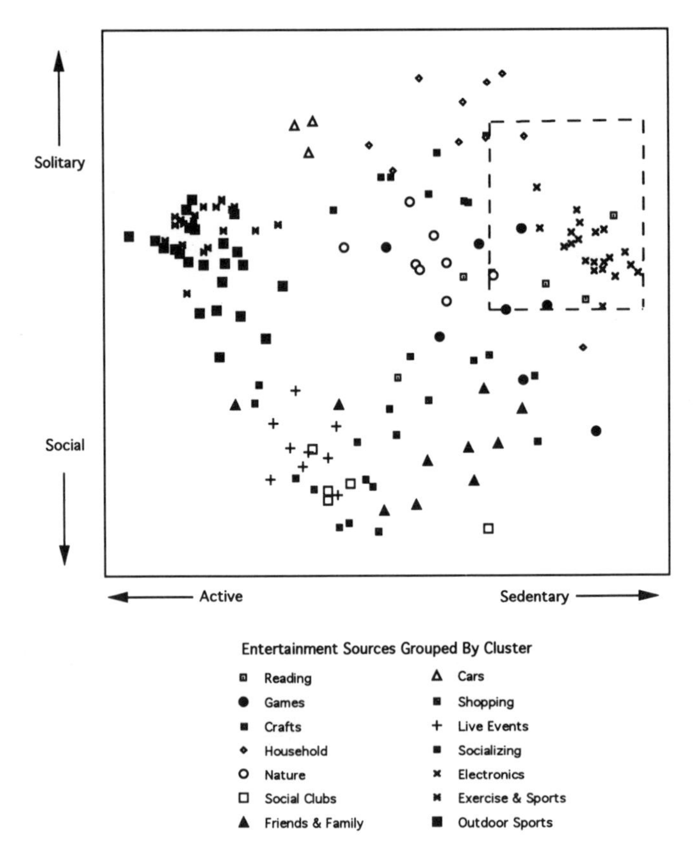

FIG. 7.2. Nonmetric MDS map of 149 entertainment sources based on judged substitutability. Each entertainment source is represented by a symbol denoting its cluster (see key). Dashed lines enclose a set of information-intensive entertainment sources.

good substitutes for each other. The home-based entertainment clusters are located together (top), as are the family and community activities (bottom, right), and the evening and weekend social activities (bottom, left).

The overall structure of the map distinguishes active and participation entertainments (left) from quiet or spectator entertainments (right), and distinguishes solitary entertainments (top) from social entertainments (bottom).[11] A possible third dimension, not plotted, distinguishes the *household* cluster from *live events* and *cars,* entertainments that emphasize travel.

[11]The coordinates in an MDS map (Fig. 7.2 and 7.3) are not supplied beforehand by the data analyst, but instead are interpretations after the fact: The original data on which the map is based do not include coordinates, but rather are information about the pairwise proximity of products, in this case the judged substitutability of pairs of products.

The section of the map containing most conventional information sources is to the far right inside the dashed rectangle, where items are labeled *reading, electronics,* and *games.* The nearest neighbors for the information-intensive sources are *crafts, household* (e.g., cooking, gardening), *nature,* and *shopping,* relatively quiet pastimes that are more solitary than social. Figure 7.3 shows detail of the section of Fig. 7.2 that is enclosed in the dashed rectangle. At upper left in Fig. 7.3 is information in the service of the household, *recipe collecting.* At lower right are electronic media, *car radio and tapes,* and *videotapes and VCR.* Between these extremes, the electronic media tend to be tightly clustered to the right, whereas print media tend to be on the periphery of this detailed section of the map. The dimensions suggested in Fig. 7.3 distinguish electronic from nonelectronic entertainment sources and forms of entertainment with a practical outcome (hobbies) from entertainments without a practical outcome (pure entertainment).

Note that entertainments employing unusual "information media," such

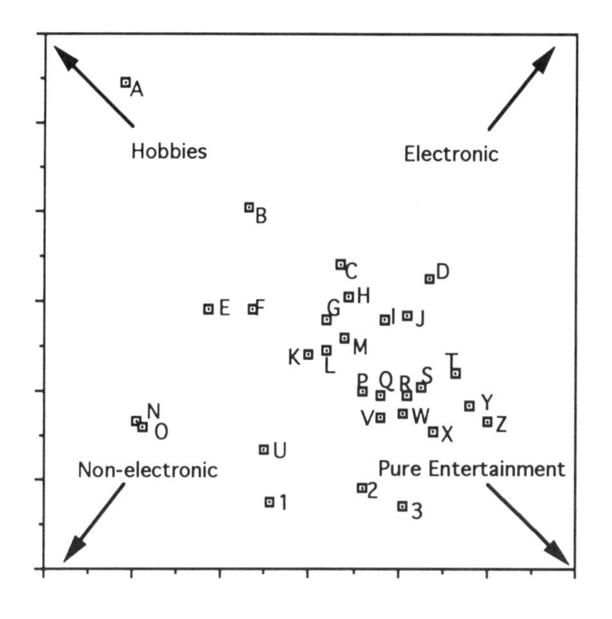

A. Recipe-collecting
B. Setting up electronics
C. Walkman
D. Books on audio tape
E. Crossword puzzles
F. Computer games
G. Video camera
H. Radio news
I. Home computer
J. Audio tapes
K. Nintendo
L. Cable sports channel
M. Radio talk shows
N. Reading
O. Attending live shows
P. CB radio
Q. Network TV
R. Games on VCR
S. Lottery
T. Video tape specialty
U. Playing instrument
V. Public TV
W. Pay-per-view
X. Stereo
Y. VCR taping movies
Z. Car radio & tapes
1. Bridge
2. Newspapers
3. Video tapes & VCR

FIG. 7.3. Detail of MDS map featuring the 29 information-intensive entertainments shown within the dashed lines in Fig. 7.2.

as *playing an instrument, lottery,* and *bridge,* are included in this map by virtue of their role as functional substitutes for other entertainments. This map suggests that new information technologies may well compete with cards and violins, as well as the more obvious existing products such as newspapers, computers, and recipe file boxes. Entertainments that do not fall in this section of the map include some that are information-intensive, but that require travel: *museum, library, bookstore browsing,* and *movies at theaters.* Some of the attraction of these entertainments may simply be that they are outside the home and may serve a more social function.

Entertainment sources with a large communication component are not located with the solitary entertainments. *Telephone conversations,* although involving electronic hardware, is located with the *socializing* cluster in the lower center of Fig. 7.2. A new information technology, like the screen phone, might well substitute for out-of-home entertainment sources like shopping, or substitute for some forms of socializing given one kind of positioning. But, given another positioning, the screen phone might substitute for electronic entertainment sources located in the dashed rectangle of Fig. 7.2.

DISCUSSION

Implications for New Information Technology Maps

New technologies may be faster, more convenient, more powerful, and less costly than existing technologies, but in the near term they can be expected to serve mainly the same consumer needs as old information technologies. To the extent that new technologies differ from the old in degree, rather than in kind, of function, they will most likely compete with the electronic and print media shown in the present maps (Fig. 7.2 and 7.3). Thus, the present maps are most applicable to near-term technologies. An example is the screen phone that can be used in communications, transactions, and information display and storage. The advantages of the screen phone in each of these domains are in ease of operation and in speed and quantity of information displayed. However, the services envisioned for this new piece of equipment are qualitatively the same as existing services offered over standard telephones (or offered outside the telephone network altogether). Thus, for a new technology such as the screen phone, present product maps should be adequate.

However, to the extent that new technologies foster new functions, the new technologies will not be represented in the present map. If new technologies are used by consumers to deliver qualitatively different kinds of information, to perform transactions differently, or to create new kinds of social interactions, then a new map may be needed. For example, teenagers might invent a new multiperson game for the screen phone that might be unlike anything in the map in Fig. 7.2. The speed with which new functions arise will depend not only on the speed of developing new technology, but also on the speed with which consumers invent new uses for information and change old habits. Thus, when new technologies threaten to change the way in which consumers seek and use information, a product manager could commission a study for new product maps based explicitly on information products. The methods demonstrated in the present study provide a relatively simple, powerful, and cost-effective means for creating such a map customized for information products and sources.

Implications for Product Development With New Information Technologies

Direct consumer judgment of product substitutability leads to a map that (a) captures the important element of substitution-in-use (Day et al., 1979), (b) includes enough products to show the broad market context, (c) exhibits good internal consistency, and (d) is relatively immune to effects of demographic differences among respondents. For the planner of new information products such a map has great value in clarifying the structure of the market into which the potential new products will be introduced. Gaps in existing product offerings are revealed and potential competitive relationships are exposed. Because of the technical robustness of the approach, managers can feel relatively confident that new concepts emerging from such a study actually reflect the market context.

The maps shown here indicate that consumers judge many forms of entertainment and information sources to be substitutable. From the consumer's point of view, substitutability depends on usage situations that are only loosely associated with technology or medium. Thus, the substitutability of entertainment sources depends more on whether they are active or sedentary, solitary or social, indoor or outdoor, than on the particular delivery medium (Fig. 7.2). The information-intensive entertainment sources are concentrated in the sedentary and more solitary part of Fig. 7.2 (dashed rectangle). Because of the concentration of products within the dashed rectangle, one might expect new information products to face stiff competition if positioned as primarily entertainment to be enjoyed alone. New information technologies might face less concentrated competition if they served functions that either (a) were targeted less for entertainment and more for hobbies and self-improvement (more toward the center of Fig. 7.2), or (b) were targeted less for solitary activity and more for social activity (more toward the lower part of Fig. 7.2).

The section on previous maps claims that product maps can be used for

(a) development of new information products, (b) choice among new product concepts, and (c) positioning. As an example, suppose a product manager were developing various packages of services to offer over a screen phone. Referring to the map in Fig. 7.2, the product manager might emphasize telephony services in order to place the product more toward the *socializing* region of the map (lower center), and de-emphasize games in order to avoid the *electronics* and *games* region of the map. Or, the product manager might emphasize development of home improvement or education services in order to place the product more toward the upper center of the map.

Suppose the product manager had budget to test just two different screen phone concepts yet had to choose among a telephony concept, a home improvement concept, and an education concept. Using the map in Fig. 7.2, the product manager realizes that the home improvement service and the education service are more substitutable for each other than either is for the telephony service. The product manager might try to maximize the amount of information that the concept test provides by choosing two very dissimilar concepts for testing. This strategy would lead to the telephony concept and one of the other two concepts being tested.

Later, when developing advertising, the product manager could consult the map to help position the screen phone product with respect to existing products. For example, suppose the product were recipes on demand delivered to a screen phone's speaker and screen. Similar products and pastimes appear in the upper center of the map in Fig. 7.2 in the clusters *household* and *crafts.* Nearby in the map are the entertainment clusters *reading* and *electronics.* The product manager might seek advertising to differentiate the product from books and paper files, on the one hand, and also from televised cooking shows. Ad tests could be designed to show whether the advertising did in fact differentiate the screen phone product from its substitutes.

Beyond the specific uses for a product map, perhaps its most important function is as a unifying metaphor. Whether for developing product features, for choosing among concepts, or for positioning, the product manager can look at the map and ask the same questions: "Where is my product? Where are my competitors? Where do I want my product to be?"

APPENDIX

Agreement With Previous Entertainment Maps

The present results from clustering and MDS analyses of direct judgments of substitutability agree most closely with previous results also based on

MDS analyses of proximity judgments. Ritchie's (1975) scaling of similarity judgments produced dimensions of active versus passive and individual versus group, as in Fig. 7.2. In addition, Ritchie identified a participant versus spectator dimension similar to the horizontal dimension in Fig. 7.2, and a home versus outside the home dimension like the possible third dimension of the MDS solution. Becker's (1976) MDS analysis of similarity judgments of discretionary activities also showed a solitary versus social dimension and an active versus sedentary dimension (Becker's Dimensions 2 and 3 collapsed). The active versus sedentary dimension also appears in studies based on a factoring or scaling of respondents-by-activities matrices (Bishop, 1970; Duncan, 1978; Holbrook & Lehmann, 1981).

The present results differ from those based on respondents-by-activities data by not revealing dimensions that reflect population segments such as low-brow versus high-brow (Hirschman, 1985; Holbrook & Lehmann, 1981; London et al., 1977; McKechnie, 1974), or male versus female (Hirschman, 1985; McKechnie, 1974). Respondents-by-activities data simultaneously contain information about the products (activities) and the respondents. Resulting map dimensions indicate between-respondents sources of variation as well as between-products sources, much as in Carroll's (1972, p. 128) "internal analysis" of preference data.

The present study, in contrast, decoupled information about the products from information about the respondents. Both the judgment task and the type of data aggregation contributed to this decoupling. The task implicitly emphasized usage situations, that is, the time, place, and purpose slot that an activity might occupy. These situations exist independently of any individual respondent's usage level or preference. Thus, different respondents would be likely to agree that activities that take place at home during the day mainly on weekends are at least somewhat substitutable, whereas the same respondents might differ in their actual usage of such activities. Thus, the substitutability judgment would tend not to distinguish respondents, whereas usage behavior would tend to distinguish respondents from each other.

Consider the case of the low-brow versus high-brow dimension, specifically *playing cards* and *going to the library*. In Holbrook and Lehmann's (1981) study, these activities anchored the ends of a dimension. In the present data, these activities were of intermediate distance from each other, whereas *sports* and *TV* anchored the dimension (Fig. 7.2). In the substitutability task, respondents were encouraged to suppose that the card game were unavailable, which implies that there is an evening indoor activity slot to fill. If respondents implicitly ordered the other activities by their suitability for filling this slot, *library* would score higher than, say, *playing football,* even if the respondent actually does play football and actually does not go to the library. In the present results, *library* (*reading* cluster,

Fig. 7.2) and *cards* (*games* cluster, Fig. 7.2) appear as more substitutable with each other than either is with *playing football* (*outdoor sports* cluster, Fig. 7.2).

ACKNOWLEDGMENT

I thank Sandra Teare for assistance in collecting data and preparing figures.

REFERENCES

Allenby, G. M. (1989). A unified approach to identifying, estimating and testing demand structures with aggregate scanner data. *Marketing Science, 8,* 265–280.

Anderson, A. B. (1973). Brief report: Additional data on distortion due to aggregation in nonmetric multidimensional scaling. *Multivariate Behavioral Research, 8,* 519–522.

Arabie, P., Carroll, J. D., & DeSarbo, W. S. (1987). *Three-way scaling and clustering.* Newbury Park, CA: Sage.

Arabie, P,. & Hubert, L. J. (1992). Combinatorial data analysis. *Annual Review of Psychology, 43,* 169–203.

Arabie, P., & Soli, S. D. (1982). The interface between the types of regression and methods of collecting proximity data. In R. G. Golledge & J. N. Rayner (Eds.), *Proximity and preference: Problems in the multidimensional analysis of large data sets* (pp. 90–115). Minneapolis: University of Minnesota Press.

Becker, B. W. (1976). Perceived similarities among recreational activities. *Journal of Leisure Research, 8,* 112–122.

Bishop, D. W. (1970). Stability of the factor structure of leisure behavior: Analyses of four communities. *Journal of Leisure Research, 2,* 160–170.

Bucklin, R. E., & Srinivasan, V. (1991). Determining inter-brand substitutability through survey measurement of consumer preference structures. *Journal of Marketing Research, 28,* 58–71.

Carroll, J. D. (1972). Individual differences and multidimensional scaling. In R. N. Shepard, A. K. Romney, & S. B. Nerlove (Eds.), *Multidimensional scaling* (Vol I/Theory) (pp. 105–155). New York: Seminar Press.

Carroll, J. D., & Arabie, P. (1980). Multidimensional scaling. *Annual Review of Psychology, 31,* 607–649.

Carroll, J. D., & Arabie, P. (1983). INDCLUS: An individual differences generalization of the ADCLUS model and the MAPCLUS algorithm. *Psychometrika, 48,* 157–169.

Carroll, J. D., & De Soete, G. (1991). Toward a new paradigm for the study of multiattribute choice behavior. *American Psychologist, 46,* 342–351.

Cooper, L. G. (1983). A review of multidimensional scaling in marketing research. *Applied Psychological Measurement, 7,* 427–450.

Day, G. S., Deutscher, T., & Ryans, A. B. (1976). Data quality, level of aggregation, and nonmetric multidimensional scaling solutions. *Journal of Marketing Research, 13,* 92–97.

Day, G. S., Shocker, A. D., & Srivastava, R. K. (1979). Customer-oriented approaches to identifying product-markets. *Journal of Marketing, 43,* 8–19.

De Leeuw, J., & Stoop, I. (1984). Upper bounds for Kruskal's stress. *Psychometrika, 49,* 391–402.

DeSarbo, W. S., & Hoffman, D. L. (1987). Constructing MDS joint spaces from binary choice data: A multidimensional unfolding threshold model for marketing research. *Journal of Marketing Research, 24,* 40–54.

DeSarbo, W. S., Jedidi, K., & Johnson, M. D. (1991). A new clustering methodology for the analysis of sorted or categorized stimuli. *Marketing Letters, 2,* 267–279.

Drasgow, F., & Jones, L. E. (1979). Multidimensional scaling of derived dissimilarities. *Multivariate Behavioral Research, 14,* 227–244.

Duncan, D. J. (1978). Leisure types: Factor analyses of leisure profiles. *Journal of Leisure Research, 10,* 113–125.

Fraser, C., & Bradford, J. W. (1983). Competitive market structure analysis: Principal partitioning of revealed substitutabilities. *Journal of Consumer Research, 10,* 15–30.

Green, P. E. (1975). Marketing applications of MDS: Assessment and outlook. *Journal of Marketing, 39,* 24–31.

Green, P. E., Carmone, F. J., & Fox, L. B. (1969). Television programme similarities: An application of subjective clustering. *Journal of the Market Research Society, 11,* 70–90.

Grover, R., & Dillon, W. R. (1985). A probablistic model for testing hypothesized hierarchical market structures. *Marketing Science, 4,* 312–335.

Hirschman, E. C. (1985). A multidimensional analysis of content preferences for leisure-time media. *Journal of Leisure Research, 17,* 14–28.

Holbrook, M. B., & Lehmann, D. R. (1981). Allocating discretionary time: Complementarity among activities. *Journal of Consumer Research, 7,* 395–406.

Jain, K. (1993). *Consumers' prior knowledge and the new media: Implications from a categorization perspective.* Unpublished working paper, Marketing Department, College of Business Administration, University of Rhode Island, Kingston.

Krishnamurthi, L., & Raj, S. P. (1991). An empirical analysis of the relationship between brand loyalty and consumer price elasticity. *Marketing Science, 10,* 172–183.

Lehmann, D. R. (1972). Judged similarity and brand-switching data as similarity measures. *Journal of Marketing Research, 9,* 331–334.

London, M., Crandall, R., & Fitzgibbons, D. (1977). The psychological structure of leisure: Activities, needs, people. *Journal of Leisure Research, 9,* 252–263.

McKechnie, G. E. (1974). The psychological structure of leisure: Past behavior. *Journal of Leisure Research, 6,* 27–45.

Miller, G. A. (1967). Psycholinguistic approaches to the study of communication. In D. L. Arm (Ed.), *Journeys in science: Small steps—great strides.* Albuquerque: University of New Mexico Press.

Miller, G. A. (1969). A psychological method to investigate verbal concepts. *Journal of Mathematical Psychology, 6,* 169–191.

Pung, G., & Stewart, D. W. (1983). Cluster analysis in marketing research: Review and suggestions for application. *Journal of Marketing Research, 20,* 134–148.

Rao, V. R., & Katz, R. (1971). Alternative multidimensional scaling methods for large stimulus sets. *Journal of Marketing Research, 8,* 488–494.

Ratneshwar, S., & Shocker, A. D. (1991). Substitution in use and the role of usage context in product category structures. *Journal of Marketing Research, 28,* 281–295.

Reynolds, T. J., & Gutman, J. (1988). Laddering theory, method, analysis, and interpretation. *Journal of Advertising Research, 28,* 11–31.

Ritchie, J. R. B. (1975). On the derivation of leisure activity types—A perceptual mapping approach. *Journal of Leisure Research, 7,* 128–140.

Rosenberg, S. (1982). The method of sorting in multivariate research with applications selected from cognitive psychology and person perception. In N. Hirschberg & L. G. Humphreys (Eds.), *Multivariate applications in the social sciences* (pp. 117–142). Hillsdale, NJ: Lawrence Erlbaum Associates.

Rosenberg, S., Nelson, C., & Vivekananthan, P. S. (1968). A multidimensional approach to the structure of personality impressions. *Journal of Personality and Social Psychology, 9,* 283–294.

SAS Institute, Inc. (1985). *SAS user's guide: Statistics, version 5 edition.* Cary, NC: Author.

Schultz, J. V., & Hubert, L. (1976). A nonparametric test for the correspondence between two proximity matrices. *Journal of Educational Statistics, 1,* 59–67.

Shepard, R. N. (1972a). Introduction to volume I. In R. N. Shepard, A. K. Romney, & S. B. Nerlove (Eds.), *Multidimensional scaling (Vol. I/Theory)* (pp. 1–20). New York: Seminar Press.

Shepard, R. N. (1972b). A taxonomy of some principal types of data and of multidimensional methods for their analysis. In R. N. Shepard, A. K. Romney, & S. B. Nerlove (Eds.), *Multidimensional scaling (Vol. I/Theory)* (pp. 21–47). New York: Seminar Press.

Shepard, R. N. (1987). Toward a universal law of generalization for psychological science. *Science, 237,* 1317–1323.

Shocker, A. D., & Srinivasan, V. (1979). Multiattribute approaches for product concept evaluation and generation: A critical review. *Journal of Marketing Research, 16,* 159–180.

Simonson, I. (1990). The effect of purchase quantity and timing on variety-seeking behavior. *Journal of Marketing Research, 27,* 150–162.

Srivastava, R. K., Alpert, M. I., & Shocker, A. D. (1984). A customer-oriented approach for determining market structures. *Journal of Marketing, 48,* 32–45.

Stefflre, V. (1986). *Developing and implementing marketing strategies.* New York: Praeger.

Stringer, P. (1967). Cluster analysis of non-verbal judgments of facial expressions. *The British Journal of Mathematical and Statistical Psychology, 20,* 71–79.

Walters, R. G. (1991). Assessing the impact of retail price promotions on product substitution, complementary purchase, and interstore sales displacement. *Journal of Marketing, 55,* 17–28.

Wilkinson, L. (1991). *SYSTAT: The system for statistics.* Evanston, IL: SYSTAT.

CHAPTER 8
Time and Technology: The Growing Nexus

Carol Felker Kaufman
Rutgers University

Paul M. Lane
Western Michigan University

PROLOGUE

Right now you may be sitting down to this chapter, in your office, at home, while working, or in any of several possible ways. Now move ahead to the future; you may be sitting down to this book printed in hardcopy, or perhaps reviewing the process of writing this chapter on a CDI system, or even interacting with a network of researchers, furthering the discussion through your computer screen, or on one of a thousand future cable channels. Such alternatives illustrate the innovative possibilities that technology brings to life. Information previously presented in static form, bounded in time, can potentially continue to generate ideas in a dynamic format, which changes with time. The possibilities are endless; identifying which products are needed, desired, and marketable is a challenge for product development teams to confront.

In addition to possibilities, technology adds challenges as well, challenges to maintain a desired quality of life while improving the possibilities which people can produce. Changing technologies have been both praised and criticized for creating faster, more productive lives, but also lives which do not match the natural and cultural clocks which produce relaxed, healthy lifestyles (Rifkin, 1987). Innovators are faced with the mandate to implement the information-based clocks of computer technologies, while also enabling man to transition within the natural limitations of biological clocks. This chapter addresses both the problems and opportunities that are created within the time-technology nexus.

Although the use of time and technology are common to all the scenarios just given, what is striking is the power that technology has to modify individuals' perception, use, and even effects of time. In the illustrations given here, a form of information that is generally shared in only finished form can actually be viewed in process, in a participative way. This added dimension that technology brings to time is the topic for this chapter.

INTRODUCTION

Consumers of the 1990s consistently report role overload and time poverty, despite their use of multiple time-saving products and services (Fram, 1991; Gross, 1987; Linder, 1970; Reilly, 1982; Robinson, 1991). This apparent contradiction suggests that something is missing in existing methods of new product research; people's time-related needs are not fully understood. The methods presented in this chapter illustrate that consumers can provide much information about the ways that products are used in their households, rather than simply reacting to new product attributes in an artificial testing setting. Their problems and suggestions can be studied through pictures, discussions, and descriptions, so that understanding is enhanced. This approach is termed *contextual inquiry.*

Contextual inquiry allows researchers and consumers alike to break away from the traditional constraints of research focusing on a specific product. Instead, needs are studied in the context of the home; thus, problems with products can be linked with the context in which they are used. Many of the benefits people describe appear related to technology's effects on their time use, and how their desired time use can affect which products are chosen. In this chapter a set of tools is proposed for investigating the nexus, or linkage between time and technology. We present the procedures that were developed to study this relationship, and present the results of a pilot study that was implemented in the Midwest in summer 1992.

THE TIME AND TECHNOLOGY NEXUS

Trade-Offs in the Home

Researchers have conducted numerous studies that attempt to model, examine, and understand how consumers meet their needs through their household's resources, the technologies that they own and acquire, and the services they purchase. Economic analysts such as Gary Becker, and Home Economics experts Ivan Beutler and Alma Owen, among others, have viewed the household as a "small factory," in which consumers attempt to produce desired satisfactions through their selection and use of the products that are available to them.

That approach, grouped under the heading of "production function studies," views the inner workings of the household as similar to that of a factory, in which scarce resources are allocated carefully toward desired goals. Such investigations can be linked with the time and motion studies of Galbraith and the early housekeeping sociology of Anne Oakley. These examinations tended to focus on whether the household members produced commodities like home-cooked meals versus purchasing them ready-made, and have not emphasized how the consumers' varied perceptions and uses of time can impact the ways in which they adopt technologies in their everyday lives (Bellante & Foster, 1984; Nichols & Fox, 1983; Peskin, 1982; Strober & Weinberg, 1980).

My Time Is Not Your Time

The variations in time perceptions among people have long been absent from product development and innovation studies in marketing and other related business disciplines. Traditional assumptions do not generally account for the differences in how people experience and use time. Instead, time has been thought to be objectively measured in standard minutes and hours (Baumol, 1973; Becker, 1965; Lancaster, 1966). This "Newtonian" view assumes that time moves at the same speed for all persons, is valued similarly, and is allocated in much the same ways by all persons. Likewise, it assumes that faster is always "better," that all people want to reduce their time spent in certain activities, and that people are most likely to do one thing at a time (Arndt, Gronmo, & Hawes, 1981; Bellante & Foster, 1984; Family Time Use, 1981; Gronau, 1977; Hefferan, 1982).

However, other researchers have instead assumed that time is in some ways unique for everyone, and the way it is experienced varies with factors such as one's personality, age, and culture. Based on the facets of subjective time, contrary to the Newtonian view, slower may be "better" for some things, people may want to reduce time in some activities and increase it in others, and individuals are found to combine all sorts of activities based on their needs and preferences. The nuances of such "subjective time" are thought to be critically relevant to the perception and use of new information technologies. Important differences in time are found in man's perception of duration and speed (Coleman, 1986; Fraser, 1981; Hirschman, 1987; Levine, 1987, 1988; Settle, Alreck, & Glasheen, 1978); and in culturally based and age-related views of past, present, and future (Gentry, Ko, & Stoltman, 1991; Graham, 1981; Hall, 1959; Lane & Kaufman, 1993; Levine, 1988; Zerubavel, 1982). Pace, synchronization, and coordination of activities also differ across cultures (Levine 1987); just as biological clocks and performance, alertness, and perception differ across individuals (Coleman, 1986; Hamner, 1981; Hoagland, 1981). Moreover, patterns of activity combinations, sequencing, and polychronic versus monochronic time use are found both in the home and in the workplace (Bluedorn, Kaufman, & Lane, 1992; Hall, 1959, 1983; Kaufman, Lane, and Lindquist, 1991).

Time Links With Technology

Technology has changed the ways in which tasks can be carried out in the home as well as in the workplace. Through new products and new technologies, tasks may take less time to accomplish, or perhaps be carried out with less skills than previously needed. For example, the microwave oven has greatly reduced the planning time and the cooking time necessary to prepare meals, as defrosting and cooking can now be completed in a matter of minutes. Likewise, offices and factories benefit from computerized technology for product manufacturing, standardizing engineering skills with great precision. Persons are able to do things faster, more accurately, with different levels of skill, through the ways that technology changes the inputs needed for them to accomplish their goals.

A major advancement in the last several decades has been vast improvements in the methods, speed, and accuracy with which information is brought into our homes and our workplaces. However, critics have challenged that many of the "advantages" of rapid technologies are actually disadvantages, and urge product developers to more closely understand the match between information needs and personal needs, as part of the context of consumers' lives.

The Slow Fax and Other Technologically Induced Perceptions

In his book *Time Wars,* Rifkin (1987) described the "new nanosecond culture," and cautions that increased efficiency may provide short-term benefits, as well as long-term psychic and environmental damage. The computer and other information technologies, he argued, are changing the way that people think about time, accelerating their sense of time, creating unmeetable deadlines and the drive to perform faster and more efficiently. Information changes faster than it can be processed, whereas depersonalized pressure separates man from the pulses of nature, in an artificial time. Functions that used to be measured in meaningful units, like minutes, hours, and days, are now occurring in imperceptible blinks of an eye, and consumers are found to observe rather contradictorily that the "fax is slow," "the e-mail took longer than a few seconds," and "the cellular phone lines are crowded."

Social, professional, and informational contexts are largely based on information technologies that transmit information faster than people can think (Simons, 1985), across spaces that were formerly impossible. Thus, the system of trade-offs that were conceptualized by early household researchers are largely based on an era in which time was perceived in clock time. Computer psychologist Brod (1984) found that technologies block out subjective time, linked to human perception, and instead anchors people's perceptions to speed, interactivity, and the instantaneous result of action. Research for the future must reflect such changes in consumers' experiences.

Technological Time Versus Clock Time Versus Subjective Time

An organizing framework for considering this dilemma is built on the comparisons and contrasts among clock time, subjective time, and technological time. Although the clock time to obtain or send information may be shortened, the time reduction may subjectively seem much shorter or only a little shorter. When activities can be combined due to technological advancement, such as driving and using the car phone, technological time allows us to combine several activities into the same block of clock time, even more dramatically altering our perceptions of the time that has elapsed.

Thus, new information technologies allow us to send, receive, and process information faster, as well as letting us manipulate many kinds of information simultaneously. Such simultaneous activity is called *polychronic time use;* and in an information context, it simply means that consumers can intake, process, and output several types of cues, signals, and data in the same time that they formerly used to process only one type of information. Electronic mail, faxes, and cellular phones enable consumers to send and receive information faster than before, and also from the previously impossible locations of their home offices, their cars, and their seat in airplanes. Shopping can be done via the television, while other programs are being recorded for future viewing, shifting one time use to another available clock time.

Figure 8.1, the technology nexus with time (TNT) matrix, illustrates several possible combinations of time and sensory use, which are possible through technologies at work and at home. Note that the time–space relationship can change as well, as home and work are linked through computer, modem, phone, and fax. Information can be exchanged impersonally, sent through modems, via faxes, and across television screens. Formerly, time-based events had to occur in the physical context in which they actually happened. Homes were warmed when consumers adjusted their thermostats, information was spoken, shopping was done in the store. Teleconferencing, electronic mail, cellular phones, and families armed with beepers are all indications that there is less need for face-to-face interaction, although perhaps a more critical need to exchange information.

The speed of perception also enables the consumer to modify their perception and use of time, relative to information. Time- compressed

TIME USE

SENSORY USE	Monochronic	Polychronic
Limited or Single Sense	**Single Sense Single Idea** Look at computer screen Watch T.V. program Q1	**Single Sense Multiple Ideas** Scan multiple Windows applications Scan multiple T.V. programs (PIP) Q2
Multiple Senses	**Multiple Senses Single Idea** Look at computer and listen to C.D. Rom on same topic Look at T.V. for Nintendo and listen to Nintendo Q4	**Multiple Senses Multiple Ideas** Scan multiple Windows applications and talk on telephone Scan multiple T.V. programs and talk on the telephone Q3

FIG. 8.1. TNT matrix.

consumers report fast-forwarding through television shows in a fraction of their intended viewing time. Voice mail is heard at accelerated rates using voice compression, eliminating the "silence" between words. Interactive searching eliminates time spent browsing through library shelves, while faxes allow us to transmit information in the blink of an eye. The other side of this picture, however, challenges if there is distortion in the fast-forwarding, information in the "silence," thought within the browsing, and value in face-to-face interaction. The balance is yet to be determined (Kaufman & Lane, 1990).

An Application to Polychronic Time Use

The notion of polychronic time use is much more complex than the simple notion of "combining more than one activity in the same clock block," but instead also includes variations in the type and amount of sensory, mental, and physical inputs devoted to each activity. Moreover, technologies are capable of enhancing the processing that is possible, unlocking new potentials for consumers' abilities. The TNT matrix illustrates some of these variations, as follows:

Q1—Individual uses a limited or single sense on a single idea or activity.
Q2—Individual uses a limited or single sense on more than one idea or activity.
Q3—Individual uses multiple senses on more than one idea or activity.
Q4—Individual uses multiple senses on one idea or activity.

The TNT matrix can be used to illustrate how individuals and families often combine the uses of several products in their homes and workplaces on a routine basis. Considering the examples in Fig. 8.1, television use may range from concentrated viewing as a single activity through scanning several programs while talking on the telephone. This matrix depicts the integration of time use and sensory use. That is, time use can focus on a single idea ("monochronic") or on multiple ideas ("polychronic"), and sensory use can range from the use of a single sense through the use of multiple senses. The range of possibilities is vast.

The TNT matrix may be a powerful tool that enables us to pull apart the various benefits that people want in terms of their actual physical and mental abilities, and the power of technology to provide abilities for them. Special attention can be given to designing products for those whose sensory capabilities need to be supplemented, such as the design for the elderly, sick, or the physically or mentally challenged.

The TIMES Model Approach: Integrating Consumer Resources for New Technology Development

In order to understand the nexus of technology and time it is necessary to also look at the relationship of time to the other resources that are available. The TIMES model is one of several taxonomies of resources (Lane & Kaufman, 1994), linked to the economics literature, which establishes an approach in which individuals use, store, and allocated their resources in their household systems throughout daily life. Here the acronym "TIMES" stands for time, information, money, energy, and space, the primary resources that household members are likely to exchange.

The TIMES model, depicted in Fig. 8.2, suggests that individuals and households have several types of resources that can be used in any given task, and that these resources are interactive and can be traded off in the way that the task is carried out. Technology, in particular, has significantly affected the types of amounts of resources that are needed to accomplish certain tasks. Note that there can be other possible configurations of resources; this model is based on those that are commonly integrated in the economics and marketing literature.

It is thought that the resources are interactive and interrelated, in that they can be exchanged, substituted, or act as complements for each other. Generally, such interrelationships depend on the individual's or collective

Time

The total clock or calendar space available to an individual or household. Recent technological products like cellular phones, have changed the amount or type of time needed for a specific action.

Information

This is knowledge or access to knowledge resources. Information is available through many services via telecommunications technology, interactive systems or fax.

Money

The financial resources that are available for use and manipulation in a variety of markets. Technology, such as computer modems and ATM machines, are changing the access to these resources.

Energy

Various levels of mental and physical energy are required to carry out an activity. New information technology, like electronic mail, can change the amount of intensity that is needed.

Space

The physical room in which one has to operate including equipment room, such as disk storage space. Since the Japanese have more limited space, their electronics easily fit in U.S. homes but the reverse is not true.

FIG. 8.2. TIMES applied to technology.

household members' preferences, abilities, and skills in using these resources. One of the things that technology does is change the relationship between these resources for an individual, a household or an organization. Examples of this include: retrieval of information electronically, whether through electronic mail, or from services such as Compuserve and Prodigy; providing time flexibility and transfer; reduction of the energy investment; the enhancement of information, reduces the space needs over hard copy processes, and usually changes the cost structure. One way to think of the information superhighway is more timely access to more information, using less energy, less money, and less space. In short much of what the telecommunications industry is working towards are changes in the relationships of these resources, in simple situations like the voice activated remote, to complex ideas like the information highway.

CONTEXTUAL INQUIRY: THE TIMES MODEL APPLIED TO THE HOME

Industries of the future have tremendous investment in understanding what problems new products are expected to solve. Specifically, it can be argued that new product developments in telecommunications are aimed at increasing individuals' time efficiencies in their home, possibly by combining familiar technologies in new ways. Developing the "right" new products, however, demands that designers understand the realities of consumers' time use, problems, and needs.

In actual study, consumers may artificially elaborate on their time use and give socially accepted answers to researchers' inquiries. Additionally, they may have difficulties in suggesting new products that can answer their needs. For instance, consumers of the 1970s may not have been likely to suggest car phones, although they could articulate the need to communicate while traveling.

In response to this problem, contextual inquiry is proposed as a new research process, which may enable management to better understand what types of new products are needed by consumers. It begins with the premise that we need to understand consumers' needs right where they occur—in the home. To develop such understanding, consumers will be asked to account for their actual time use in a diary format, also to provide pictures or sketches of the technologies that they have in their homes.

In developing this approach, an examination of methods used to study time uncovers a wide variety of formats including time logs, summary time data, surveys, drawings, focus groups, and interviews (Kaufman & Lane, 1991; Szalai, 1972; Walker & Woods, 1976). These methods are thought to be complementary, each providing an added dimension of people's lives. Such data is difficult to interpret into conclusive evidence about how households "work." Despite numerous studies on household resources, the real trade-offs are unclear, uncovering how each member chooses to "spend" their time dollars, energy dollars, monetary dollars, informational dollars, and space dollars.

A Methodology to Study TIMES: Contextual Inquiry

Contextual inquiry as applied to the current study is illustrated in Fig. 8.3, which shows how each of the consumers' resources could be studied in several ways. Fig. 8.3 is presented with the TIMES resources on the vertical axis. Across the horizontal axis are the authors' primary sources for information for each kind of resource. In the case of time, the authors found time diaries to provide the most detailed insight into a consumer's actual day, followed by survey information, and then the focus groups. Similarly, skipping down to "space," the authors found the schematics to be the most informative, followed by the focus group discussions, and then the photographs. However, all the types of data worked together to provide a much more complete understanding of each respondent, rather than more conventional research, based primarily on a survey.

This approach enables us to examine the costs and benefits of new

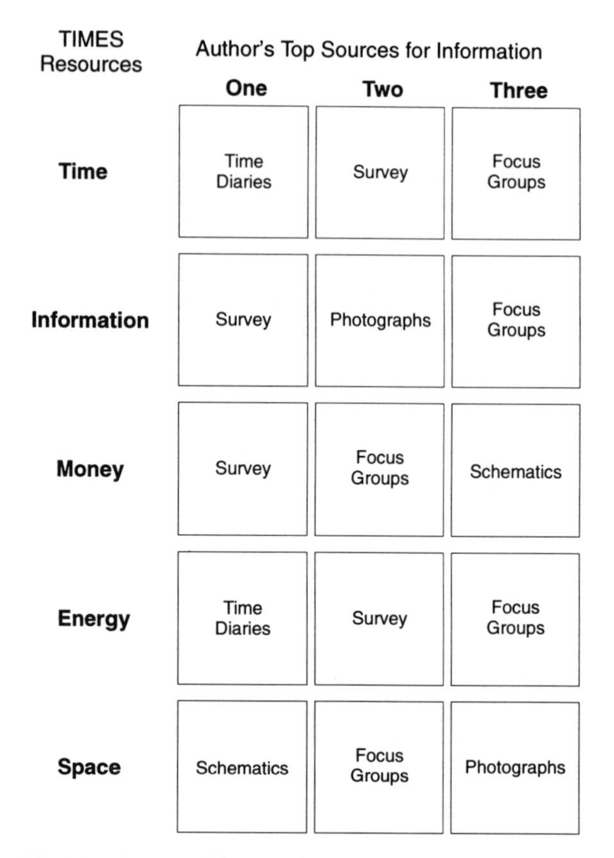

FIG. 8.3. Contextual inquiry: Integrating TIMES into research.

information technologies as part of consumers' actual household situations. The focus of our investigation were televisions, computers, and telephones within the private residences of consumers. Both televisions and telephones are fixtures in most consumers' homes, and computers are rapidly gaining such widespread acceptance. New product companies promise the integration of these three technologies into some format that provides time savings to consumers, but such combinations can be designed in many different ways. These products were selected for research in this initial study since they are on the cutting edge of new technology development.

The Sample: Multiple Methods of Data Collection

The research was conducted during summer 1992 at a campus of a midwestern university. The subjects for the study were 39 MBA students in an upper-level new product development course, along with their households, and their products. Most of the participants were employed full time in diverse "career"-type professions, ranging from engineering, sales, and banking, to financial planners and pharmacists. The majority were married, with slightly fewer females than males participating in the study. The average age was 29 years old, spanning a range of ages from 23 to 41. Incomes were widely dispersed, from a low under $20,000, through the highest over $100,000, with average household incomes between $50,000 and $60,000.

Such a specific group can potentially introduce some bias because since all are highly educated and incomes are above the national average. The sample would be expected to be more likely to have multiple televisions, phones, and computers, given the higher education and income ranges that are found. However, such characteristics can be advantageous because they increase the likelihood of uncovering patterns of multiple product use and suggest a higher than average familiarity with computers in the home and in the workplace.

At the beginning of the course, these respondents were asked to perform several types of data gathering: First, respondents were provided single-use cameras and asked to photograph their primary computers, telephones, and televisions in the home. This method was employed in order to gain accurate information regarding product locations and proximities, and also to enable the respondents to freely interpret and recall details concerning their placement in the homes (Chiozzi, 1989; Collier, 1987; Heisley & Levy, 1991). The cameras were collected, and all films were developed under uniform conditions. After development, respondents were allowed to view the photographs first, and withdraw any that they did not wish to release to the research. They next were asked to group the pictures in albums, and to provide a schematic of the insides of their homes, that identified the relative placement of each technology which was photographed. The schematics provided us with the spatial arrangements of the consumers' technologies, again in the context of all the resources possessed within the household. The subjects participated in a focus group immediately after completing their album pages and schematics, followed by a survey administration and debriefing.

During the time when the photos were being developed, each respondent was also requested to fill in a time log, or diary, in which they recorded all their activities over a 4-day block of time. This method, developed and frequently used in home economics (Barclay & Lytton, 1989; Beutler & Owen, 1980; Family Time Use, 1981; Hefferan, 1982), enables individuals to keep a record of all their activities, at home, work, and elsewhere. Our specific time log allowed for free-form entering of each activity, in that respondents were encouraged to record activities when they actually happened, and as many which were happening at the same time.

The multiple methods provided many benefits that allowed us to see the resources in action. Spatial arrangements among the products could be explored, and a richness beyond simple reports of product usage becomes possible. Although several respondents may report similar activities on time logs, such as using a home computer, the pictures and schematics allowed us to track computers in living room, kitchens, and on dining room tables; televisions stacked two and three high in bedrooms and dens; and mazes of wires, plugs, and remote controls to boggle the best innovator. Multiperson usage or proximity to some technologies are placed in context and relationships are visible. Responses in these multiple forms allowed us to understand data that would have previously been questionable, helping to bridge the gap between the product design group and the actual consumer.

A CONTENT ANALYSIS: INFORMATION TECHNOLOGIES RESEARCH

In content analyzing the multicontext data, the focus was on computers, televisions, and telephones. A trained research assistant, who had not participated in any of the earlier phases of the project, was hired to analyze the time logs, focus group summaries, surveys, photographs, and schematics as follows:

1. To investigate the consistency among the various types of data for each subject in the study, determining whether their responses were possible as reported, and to verify unexpected or surprising findings.
2. To create a one-page summary of each respondent in terms of their preferences, problems, and uses of these household information technologies.
3. To identify common sets of needs and suggestions among the respondents related to new product development in telephones, televisions, and computers.
4. To identify common sets of problems with the respondents' existing products, and suggestions to new product developers.

The content analysis identified six key concepts that are thought to characterize the benefits that consumers seek from the information technologies of the future (Klopp, 1993). These are integration and consolidation, flexibility and adaptability, convenience, mobility, efficiency, and affordability. Although there is no attempt to include all possibilities, our research suggests that the six key concepts, outlined in Table 8.1, represent a taxonomy of resource trade-offs that are critical to consumers of the

TABLE 8.1
Key Concepts for Product Design

Integration/consolidation: Products that were integrated into one unit and could be operated with one remote control, such as a fax and answering machine built into their home computers, smaller and more compact products and products actually built into the structure of the home and would hide away when not in use.
Flexibility/adaptability: Flexible and easily modified products, easy to understand and use, and offer the ability to be upgraded when new and more advanced technology is developed. Consumers recommended that changes and add-ons to computers be modified from internal adaptation to outside, easy to use ports or slots, which do not take up additional space.
Convenience: Not only reducing time to complete an activity but products that are easier to use and understand, reducing information and energy. Consumers want products that will allow them to do several activities at once or even start some activities for them when they are not at home yet, such as shop interactively via barcoded catalogue and laser pen over modem.
Mobility: Allow people to move freely throughout the home and complete several activities at once or product that can be moved easily or have their functions transferred to different parts of the house. Another area is products that can operate indoors, outdoors, in the car, or wherever they may be, expanding the range of possibilities, such as notebook computers, pocket phones, and beepers.
Efficiency: Use resources efficiently, in terms of each consumer's system of priorities. Many consumers want products that are energy efficient, will cost less to use, and will save them money and time. Products that allow them to complete activities the quickest way possible, such as using systems such as Prodigy to track information regarding airlines, stocks, and so forth.
Affordability: Affordable, justified prices of all the resources including time, energy and space, not just monetary. Consumers desire more information technologies in the kitchen, where space is at a premium, challenging them to find places for faxes and computers alongside the microwave and blender.

future. These six possible tradeoffs suggest changes in some of the resource relationships which are thought to be related to increased consumer satisfaction of unmet needs.

FINDINGS

There are a number of findings from this research, and all of them are of a qualitative nature. Our challenge was to develop a process of finding out what people will want from organizations who can manufacture or market almost anything. It is important to remember that the study was conducted to look at the telecommunications area, specifically telephones, televisions, and computers. Thus, the findings are presented in two parts: first, the focus on the telecommunications industry and second, a discussion of the research process.

Findings From the Focus on Telecommunications

The challenge of the content analysis was twofold: to determine what was learned in each individual case household and to synthesize what was learned from the households in summary. The individual case content analysis distilled the five methods of research on each study participant into approximately one page for each case. It is at this individual level that contextual inquiry allows one to look at the integration of resources and the interaction of those resources with technology; it is here that the nexus can really be seen. The next challenge was to find the commonalities between all of the cases. It is this distillation large amounts of data that led to the six key concepts identified in this study.

Figure 8.4 arrays the six key concepts, (integration, flexibility, convenience, mobility, efficiency, and affordability) against the five household resources (TIMES) in order to identify the set of resource trade-offs that appear to underlie the respondents' needs, problems, and ideas for new information technologies. The new product decoding matrix shows the key concepts on the vertical axis, and the TIMES resources across the top. The authors have attempted to show by the use of "delta" for "change" and a dark circle to represent a constant, or no change, which resources are believed to be most affected by each key concept. Consumers were working

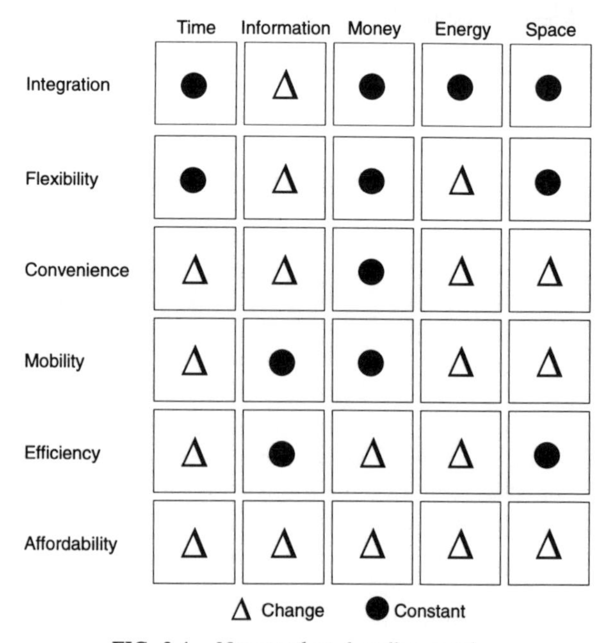

FIG. 8.4. New product decoding matrix.

toward a "balance" of resources that were optimal for them in their current lifestyles.

The study participants clearly wanted to be able to match their information needs with the products that they purchased. They experienced an increased need to integrate their technologies into their homes. They did not want to simply purchase computers, phones, and televisions, which performed general functions much the same for everyone. Instead, they preferred products with some flexibility to be customized to their individual needs. Consider the key concept "convenience": Consumers want to receive more time, information, energy, and space-saving options on their products.

They felt very strongly that they were coping with ill-designed products and home layouts that represented household activities of the "traditional" home, in which special purpose rooms were designated for sewing or laundry, while home offices were often missing. Space use seems to be quite important to the consumer of the 1990s; space convenience and mobility need to be built into products. Current products and homes quickly become obsolete, not able to keep pace with the consumers'ever-changing needs for flexibility and growth.

New homes can also build in flexibility that allows for technologies to be moved form place to place as desired by households. Numerous participants complained about "wire clutter" and predetermined locations for their household technologies that bound them to fixed power outlets and space availabilities. That is, builders' predetermined assumptions were that televisions would be in certain rooms, computers in others, with little attention paid to the mobility of actual consumers to move through their homes when the technologies are actually in use.

Today's consumers instead would like to access their business, educational, and social interest information as conveniently as possible, and they are willing to pay for it, as long as it fits their lifestyle. Instead of fitting their home offices in the corners of their bedrooms or on the dining room table, they would prefer to have readily available sources of information that are accessible through the technologies they use everyday, such as telephones, televisions, and computers. Many consumers view such information technologies as parts of their daily lives, and would prefer to have them part of the "mainstream" household layout as well.

Information technology in the home of the future may integrate key technologies as part of the structure itself. Participants in our study reported that they would prefer homes already wired and equipped with home management systems or "Smart House" capabilities. Having all clocks integrated for uniform setting, alarms, and coordination with appliances was frequently mentioned. That is, following power outages, consumers would be able to set one main timekeeping control system, rather than be faced with various banks of blinking clocks, thermostats, and timers, all with different formats for correcting the time.

Process Findings Using Contextual Inquiry

Contextual inquiry helps the researcher to explore and probe for new findings, rather than simply verifying familiar and predominant patterns. It forces the user to think about the wide variety of possible problems that consumers face, which might not be apparent to new product developers. For instance, although the subjects were not asked to specifically consider the power sources for telephones, televisions, and computers, the spider web of wires that appeared in almost every set of pictures stimulated much discussion and complaint in the subsequent focus groups. Similarly, the relative locations of such products found on the schematics gives cause for new thought. Telephones, televisions, and computers were grouped in diverse typical and unusual ways, which would not all fit the stereotypical "expected" arrangements in the home. Instead, these products were often crowded together in locations that would promise poor use, little functionality, and interference with other activities in the home.

Reality—Whose Reality? In any kind of research, it is important to capture the reality of the participants in an unbiased way. However, data can often be influenced by the collector's assumptions and expectations, especially in methods using surveys as the major source of information. This study's use of a number of collection devices seemed to stimulate frequent reality checks by the participants. The respondents knew that several types of data reflected their households, and they seemed to be highly motivated to provide a complete picture. In addition, participants had taken pictures of their household's telecommunications products prior to the focus group. Because they were able to refer back to these pictures, participants were less likely to adapt to other individual's perceptions.

Relationships: The Source of Inspiration. Contextual inquiry is designed to explore relationships among products and resources in the home, rather than focusing on single products alone. It was believed that out of the changing relationships and interactions would come the new product ideas. Contextual inquiry was successful in allowing developing new views into relationships.

Content Analysis: A Complex Process. The process of content analysis is always subjective and usually complex. The process of trying to analyze photos, schematics, focus groups, time logs, surveys, and all the interactions is detailed and tedious. The researcher must identify patterns within each person's set of multiple data, as well as identify if patterns exist overall across all those in the sample. In this research we used a somewhat homogeneous group demographically as our sample. That may be a relevant technique in contextual analysis because of the amount of information that is gathered and analyzed.

Is What They Say, What They Do? What do people want versus what they practice? These are old research problems and questions. Contextual inquiry leads to an opportunity to compare what people do and what they appear to practice, versus what they report in formats such as surveys and focus groups. The reality of consumers' lives is critical when a firm is trying to develop technologies that integrate products or services in the ways that they will actually be used.

An Idea Generation Tool: Not for Every Place and Time. Contextual inquiry seems to work well in trying to understand the needs and wants expressed and unexpressed for products of the future. Such needs could be found in schematics, pictures, and time logs, even though respondents could not put them into words in the typical research format. The concepts generated would work well for guiding idea generation, and for looking at product revisions. Such applications have contributed to product changes for a major manufacturer of household products in subsequent work done by the authors. In contrast, the use of contextual inquiry as described here would seem to be impractical and expensive for things like idea screening, concept testing, promotional planning, and so on. The intense multiple data collection would simply not be cost effective.

Competitive Advantage. Contextual inquiry develops some core ideas that may well help develop a competitive advantage for an organization. The huge knowledge base gained and the distillation of that knowledge could empower management with information and stimulate ideas that would not be available to the competition.

Other Industries. The research techniques used to conduct contextual inquiry in the exploration of the time–technology nexus were chosen for the purpose of advising the telecommunications industry. A different kind of product might suggest the use of some other combinations of techniques. There are several techniques that researchers might want to explore, such as focus groups, interviews, schematics, drawings, photographs, videotaping, time logs, tape recording, model construction, observation, house inspections, family interviews, and surveys. The list could easily be continued. Basically, the researcher should select several types of complementary research providing information representing the experiences of at the respondents in their own context.

MANAGERIAL IMPLICATIONS FOR IDEA GENERATION

There are a number of managerial implications to this exploration of time and technology and the system of contextual inquiry.

Competitive Advantage: Winning the Prize

The product development group needs to design new information technologies that provide benefits to customers. They need to know what kinds of trade-offs the customers are interested in making. This set of six concepts represents the summary set of trade-offs that this sample of respondents were interested in making in the technologies area. Technology is moving ahead so rapidly that the prize will go to those who most successfully predict the direction of the consumer wants, rather than to those who most accurately determine what today's consumer wants.

New Product Development: Customers Tell You How

The consumers offered numerous suggestions for industry that cover several possible avenues for product development:

1. improving existing products (adding functions involving physical changes to existing products, such as multiple function televisions);
2. improving existing products (adding services that enhance the capability of existing products, such as the availability of hundreds of television channels);
3. combining existing products in new ways (e.g., combining telephones and televisions in one unit);
4. innovations, which suggest new information technologies that meet consumers needs.

The Product Idea Generation Matrix

Models often appear to be simplistic. However, their simplicity can allow management to take very complex ideas, such as the nexus of time and future technologies, and attempt to array these ideas in neat classification "boxes." The pieces that do not fit well or that stand out command our attention by their differences. By applying a series of models, new information can be identified, providing management with a method of addressing the unknown in small usable chunks.

The authors propose a product idea generation (PIG) matrix to organize these types changes by the resource use which they modify. The PIG matrix illustrates the nexus of the TIMES resources and various types of product change. On the vertical axis are the avenues proposed for product development. In Fig. 8.5, the chart has been completed for illustration purposes with the results of the current study focusing on computers, telephones, and televisions.

A result that clearly stands out was the necessity to integrate household appliances with one's life, whether the consumer is at home or still at work. Participants strongly supported technology that would allow them to "call home" to communicate with their appliances. Calling from one's work to turn on the oven, adjust the thermostat, retrieve messages, and do the laundry are all examples of options that were suggested. An overwhelming interest in voice activation of home appliances while in the home was noted as well. This interest supplements the energy-savings gained from hands-free use of those appliances.

Moreover, kitchens are the information centers of busy families, however they are organized more like production factories of the economic models, oriented to the single purpose of cooking. Decisions are made, homework is done, bills are paid, and meals are prepared and eaten in kitchens across the country, thus it is not surprising that consumers would like products designed that fit all these needs.

	Improve Existing Products	**Improve Services on Existing Products**	**Combine Existing Products**	**Innovations**
Time	T.V. and other activities. Hands-free phone and other technologies.	Voice mail.	Integrated answering machine.	Solve time limitations. Portability.
Information	Display incoming call information on telephone.	Information highway. More channels. Screen calls. T.V. on demand.	Fax in kitchen. Phone/T.V. combined with home intercom.	Interactive CD. Home intercom. Installed smart cables in house.
Money	Read out on phone of time and money spent.	Pay bills-banking by phone/modem. Shop by phone, modem or fax.	Security with computer.	Security systems.
Energy	Combine remotes. Activation by voice, motion. Hands-free.	Control utilities efficiently. Automatic call-in to central control.	One remote control for all appliances T.V./VCR/CD etc.	Mobility through home, wireless, portable. Flexible home set-up.
Space	Voice activation. Speaker telephones in rooms. Wireless.	Central command to set/control all items, e.g. clocks.	Call home to communicate with appliances.	Moveable outlets. Built-in furniture which fits technology.

FIG. 8.5. PIG matrix.

SUMMARY

Industry has recognized the growing importance of consumers' experience and perception of time in the development and modification of products for the future. Time is complex, often experienced differently by some people, whereas patterns of time use can be aggregated among groups of people. Although such common patterns have often predominated in industry response to consumer needs, the unique realities of consumers' lives are essential to understand as well. Studying time in terms of product development thus requires methods that can both capture the subjective experience and the common, objectively measured uses of time that form the bases for new product needs.

Product management teams lament that consumers frequently cannot express readily their actual needs when presented with prototypes in research investigations. Such methods run the risk of abstracting products away from the contexts in which they are used. Traditional studies instead frame the consumers' reactions in terms of products isolated from the realities of the home and the way that such items are integrated into daily living.

Contextual inquiry provides a way to study the consumer in their actual situation, in order to uncover the actual problems and needs which they face, and the current ways in which they are using technologies to meet their needs. Consumers can be "observed," using and misusing their assortment of products, arranged in their own unique configurations in the home. Respondents can identify their new product needs more realistically, while reviewing their homes through cameras, schematics, time diaries, and the like. It enables product developers to analyze time use in the context of other resource trade-offs.

Other industries can uncover "hidden" information through adopting the contextual inquiry approach, when the nexus between time and their products are being studied. Multiple methods can be selected and integrated that similarly allow consumers' experiences, reactions, problems, and satisfactions to be understood.

ACKNOWLEDGMENTS

We would like to express our appreciation to Ruby Roy and Nik Dholakia of RITIM for their financial support and intellectual stimulation in the development of this research. We would like to acknowledge the many students who help us to continue to explore the linkages between time and technology, and in particular would like to thank Alyssa Klopp for detailed content analysis, Jennifer Kaufman for coding and processing surveys, Cassandra Anlage for literature integration, Ethel Cathers for in-depth study of the elderly, and Jill Anderson for the organization and administration of the contextual research. In addition, we recognize the contributions of Maria Suchowski in the focus group administration, Gary Goscenski for tables and graphics, and Malcolm V. Lane for critical editorial work.

REFERENCES

Arndt, J., Gronmo, S., & Hawes, D. (1981). The use of time as an expression of life-style: A cross-national study. *Research in Marketing, 5, 1–28.*

Barclay, N., & Lytton, R. (1989). *Household time use.* Paper presented at The Time/Quality of Life Interface, Third Quality of Life/Marketing Conference, Blacksburg, VA.

Baumol, W. (1973). Income and substitution effects in the Linder theorem. *Quarterly Journal of Economics, 87,* 629–633.

Becker, G. (1965). A theory of the allocation of time. *The Economic Journal, 75,* 493–517.

Bellante, D., & Foster, A.C. (1984). Working wives and expenditures on services. *Journal of Consumer Research, 11,* 700–707.

Beutler, I., & Owen, A. J. (1980). A home production activity model. *Home Economics Research Journal, 9,* 16–26.

Bluedorn, A. C., Kaufman, C. J. & Lane, P. M. (1992). How many things do you like to do at once? An introduction to monochronic and polychronic time. *The Academy of Management Executive, 6*(4), 17–26.

Brod, C. (1984). *Technostress.* Reading, MA: Addison-Wesley.

Chiozzi, P. (1989). Reflections on ethnographic film with a general bibliography. *Visual Anthropology, 2*(1), 1–84. Coleman, R. M. (1986). *Wide awake at 3 a.m.: By choice or by chance?* New York: Freeman and Company.

Collier, J. (1987). Visual anthropology. In J. Wagner (Ed.), *Images of information* (pp. 161–169). Beverly Hills, CA: Sage.

Family time use: An eleven-state urban/rural comparison. (1981). Blacksburg: Virginia Agricultural Experiment Station, Bulletin VPI-2.

Fram, E. (1991, Summer). The time compressed shopper. *Marketing Insights,* 34–39.

Fraser, J. T. (1981). *The voices of time.* Amherst: The University of Massachusetts Press.

Gentry, J. W., Ko, G., & Stoltman, J. J. (1991). Measures of time orientation. In J-C. Chebat & V. Venkatesan (Eds.), *Proceedings of the time and consumer behavior conference.* Val Morin: University of Quebec at Montreal.

Graham, R. J. (1981). The role of perception of time in consumer research. *Journal of Consumer Research, 7,* 335–342.

Gronau, R. (1977). Leisure, home production, and work: The theory of the allocation of time revisited. *Journal of Political Economy, 85,* 1099–1123.

Gross, B. (1987). Time scarcity: Interdisciplinary perspectives and implications for consumer behavior. In J. N. Sheth & E. C. Hirschman (Eds.), *Research in consumer behavior* (Vol. 2, pp. 1–54). Greenwich CT: JAI Press.

Hall, E. T. (1959). *The silent language.* Garden City, NY: Doubleday.

Hall, E. G. (1983). *The dance of life: The other dimension of time.* Garden City, NY: Anchor Press/Doubleday.

Hamner, K. C. (1981). Experimental evidence for the biological clock. In J. T. Fraser (Ed.), *The voices of time* (pp. 281–295). Amherst: The University of Massachusetts Press.

Hefferan, C. (1982). New methods for studying household production. In K. K. School & K. S. Tippett (Eds.), *Family economics review* (Vol. 3, pp. 30–33). Hyattsville, MD: Agricultural Research Service.

Heisley, D. D., & Levy, S. J. (1991). Autodriving: A photoelicitation technique. *Journal of Consumer Research, 18*(3), 257–272.

Hirschman, E. (1987). Theoretical perspectives of time use: implications for consumer behavior research. In J. N. Sheth & E. C. Hirschman (Eds.), *Research in consumer behavior* (Vol. 2, pp. 55–81). Greenwich, CT: JAI Press.

Hoagland, H. (1981). Some biochemical considerations of time. In J. T. Fraser (Ed.), *The voices of time* (pp. 312–379). Amherst: The University of Massachusetts Press.

Kaufman, C. J., & Lane, P. M. (1990). Quality of life in the rat race: Household management and time use as viewed from the fourth dimension. *Proceedings of the Quality of Life Conference,* 412–423.

Kaufman, C. J., & Lane, P. M. (1991). Bridging the time use measurement gap: Insights, issues, and problems from five major time use studies. *Proceedings of the 1991 Southern Marketing Association Conference,* 88–93.

Kaufman, C. J., Lane, P. M., & Lindquist, J. D. (1991). Exploring more than twenty-four hours a day: A preliminary investigation of polychronic time use. *Journal of Consumer Research, 18,* 392–401.

Klopp, A. (1993). *Preparing for the future: How research can benefit the telecommunications and housing industries.* Unpublished manuscript.

Lancaster, K. A. (1966). A new approach to consumer theory. *Journal of Political Economy, 74,* 132–157.

Lane, P. M., & Kaufman, C. F. (1993). Using time in strategic marketing. In M. J. Baker (Ed.), *Perspectives on marketing management* (Vol. 3, pp. 333–357). New York: Wiley.

Lane, P, M., & Kaufman, C. F. (1994). Finding TIMES in household technologies. In E. Wilson (Ed.), *Developments in marketing science (Vol. 17, pp. 129–132). Nashville, TN: Academy of Marketing Science, forthcoming.*

Levine, R. V. (1987). The pace of life across cultures. In J. E. McGrath (Ed.), *The social psychology of time* (pp. 39–60). Beverly Hills: Sage.

Levine, R. V. (1988, April). Waiting is a power game. *Psychology Today,* 24–33.

Linder, S. (1970). *The harried leisure class.* New York: Columbia University Press.

Nichols, S. Y., & Fox, K. D. (1983). Buying time and saving time: strategies for managing household production. *Journal of Consumer Research, 10,* 197–208.

Peskin, J. (1982). Measuring household production for the GNP. *Family Economics Review, 3,* 16–25.

Reilly, M. D. (1982). Working wives and convenience consumption. *Journal of Consumer Research, 8,* 407–418.

Rifkin, J. (1987). *Time wars: The primary conflict in human history.* New York: Touchstone Books.

Robinson, J. P. (1991, November). Your money or your time. *American Demographics,* 22–26.

Settle, R., Alreck, P., & Glasheen, J. W. (1978). Individual time orientation and consumer life style. In H. K. Hunt (Ed.), *Advances in consumer research* (Vol. 5, pp. 315–319). Ann Arbor, MI: Association for Consumer Research.

Simons, G. (1985). *Silicon shock: The menace of the computer.* Oxford: Basil Blackwell.

Strober, M. H., & Weinberg, C. B. (1980). Strategies used by working and nonworking wives to reduce time pressures. *Journal of Consumer Research, 6,* 338–348.

Szalai, A. (1972). *The use of time.* The Hague: Mouton.

Walker, K., & Woods, M. E. (1976). *Time use: A measure of household production of family goods and services.* Washington DC: American Home Economics Association.

Zerubavel, E. (1982). The standardization of time: A sociohistorical perspective. *The American Journal of Sociology, 88,* 1–23.

CHAPTER 9
Adoption of Information Technology: Contributing Factors

Norbert Mundorf
Stuart Westin
University of Rhode Island

Telecommunications providers have traditionally focused on voice-based (i.e., telephone) technologies. As the U.S., western European, and Japanese telephone markets reach a point of saturation with limited growth prospects for additional voice-based services, there is rising interest in exploring the potential for screen-based information technologies. These include videotex information services such as Prodigy, networks such as Internet, video conferencing, videophone, as well as numerous other interactive applications.

Recent discussion of the 500-channel environment (Carter, 1994) has drawn attention to electronic shopping (Strom, 1994), banking, health care, and education (Armstrong, Yang, & Cuneo, 1994) in the home. Many such applications are already technically feasible; however, it is unclear which user interface and content characteristics are conducive to attracting the average user to such services (Lohr, 1994).

Videotex and similar screen-based information services have attracted particular attention because they represent not only an improved and alternative transmission technique to traditional telephone and television, but also provide the means to link large numbers of information providers and receivers in flexible and interactive formats.

The Technology User

The climate for communication and information technologies has been changing rapidly in recent years (Antonoff, 1989; Schlossberg, 1991). For the average user, technical capabilities of communication and computer systems have by far exceeded the level of what is needed in terms of functionality ("I can't work"; 1991; Rooney, 1991). Venkatesh and Vitalari (1987) found that people tend toward simpler computer applications than they had initially planned to use.

Appealing to nonexperts is particularly important for information technologies that are marketed for home use. Such products and services are often designed to take the place of established and familiar information and communication technologies (phone, TV, newspapers, magazines) or face-to-face interaction. To many potential users, the advantage of the new over the old technology is not necessarily evident. Getting used to the new technology is often perceived as an inconvenience, which does not outweigh potential benefits.

In a competitive and fragmented communications environment such as the United States, demand-side factors play a far more significant role than they do in state-led communications environments such as France or Singapore (Sisodia, 1991). In deregulating and competitive environments, user acceptance of information technologies cannot be understood without careful study of how users respond to various communication and information formats and attributes.

The studies reported in this chapter are part of a larger program of research that is investigating the consumer acceptance of new information technologies in an international setting. Consumer acceptance and adoption of technology is a function of many different factors. From a marketing perspective, we are concerned with individual differences, notably demographic factors such as gender, age, culture, as well as characteristics of the product such as content, structure, and user interface of communication technologies.

In particular, attitudes and orientations toward technologies of U.S. and German consumers in different age groups are compared. This survey research is complemented by a series of experiments in which the effects of software and user interface features are tested. Although the surveys cover all age groups, the experimental subjects are U.S. college students, who are among the leading users of many evolving information technologies.

Demographic Factors

Gender. In the Information Age, the ability to manage information technologies is a critical condition for one's professional and private success in today's society (cf. Salvaggio & Bryant, 1989). In the corporate environment, computers, telecommunications and information networks are critical for effective job performance for an ever increasing number of people. Even outside the office electronic formats for banking, shopping, dating, magazines, elder care, and health care are gaining acceptance. Because of the importance of technological competence, possible gender bias is considered a critical issue in the use of and attitudes toward computers and related communication technologies. Some authors (Lockheed, 1985; Rogers, 1986) expressed concern that women might be at a disadvantage in these areas.

Gender differences materialize even in the use of more traditional technologies such as television. Besides age, gender is the strongest predictor of television content preferences (cf. Beville, 1988). Ferguson (1991) and others found ample evidence for a male bias in remote control use. Zillmann, Weaver, Mundorf, and Aust (1986) suggested that entertainment provides a vicarious way of reexperiencing male–female differences that have dissipated in real life. Gender-specific behavior may have lessened in the work sphere; but it is still exhibited while viewing or reading entertainment.

However, gender differences in computer use still stretch across the private and public spheres (Arch & Cummins, 1989; Kiesler, Sproull, & Eccles, 1985; Venkatesh & Vitalari, 1986, 1987). Rogers (1986) claimed that new information technologies, especially computers, have caused gender inequalities, and even reversed some of women's gains in the workplace. He argued that the male bias in math and science has extended to computers. Brownell and Mundorf (1990) reported that college males report higher usage levels of word processing, electronic mail, and statistical analysis than women. This survey confirmed findings from earlier research that men show greater use of computers, electronic mail, and similar activities.

Several authors suggested that this bias is a function of cultural and social influences as well as software design and user interface factors. Notably, the presence of well-defined structures in the computing environment has been shown to mitigate gender differences. Kiesler et al. (1985) suggested that social circumstances and software content are instrumental in perpetuating the male bias. They contended that entertaining computer software is often designed in such a way that men find it more appealing based on content (war games, contact sports) and structure (spatial cues).

Recently, information services have emerged that appear to have a relatively stronger following among women (Carey, 1991). In several cases, an adequate teaching environment and ample computing resources have led to a disappearance of this bias. It might be that such a softening of the gender bias is the result of new software content and structure that has considerable appeal to women. Miura (cited in Kubey & Larson, 1990) pointed out that girls become more involved with software applications that they find socially or intellectually meaningful, whereas boys prefer action-oriented applications. Meyer and Schulze (1993, 1994) reported that males tend to be interested in the technology for its own sake, whereas females look toward the benefits provided by the software.

In order to understand better how users relate to different technologies, it might be important to explore perceived groupings of various information and entertainment media, and gender-based familiarity with these technologies. Users may perceive a screen-based information service as more similar to the telephone or television than to a computer system. Telephone and television are used more by women (Mundorf, Meyer, Schulze, & Zoche, 1994).

Age. Older adults have had dramatically different experiences with new technologies than have younger adults. Just as many devices were becoming affordable to the average U.S. household, some older adults were retiring and facing limited incomes. After functioning well for decades without the technology, every older adult may not view a fax, or cellular phone, compact disc player as an essential tool or appliance.

As a result, older consumers tend to be more reluctant to use emerging technologies than their younger counterparts. This tendency is somewhat confounded with gender effects, as men across age groups feel more comfortable with most technologies than women (Meyer & Schulze, 1993, 1994). Some of these age differences are apparently the result of societal misconceptions. New communication and information technologies, such as electronic shopping, banking, and mail, videophone, telemedicine, and security systems may help older adults transcend environmental and social barriers (Nussbaum, Thompson, & Robinson, 1989). Unfortunately, older adults are also generally more reluctant to use such technologies (Kerschner & Chelsvig, 1981; Zeithaml & Gilly, 1987).

Culture/Nationality. A critical factor influencing the acceptance of and response to information technologies internationally is culture. Differences between consumers of information technology products and services from different cultures could arise from a variety of sources, including political and social trends, educational, environmental, legal, and historical factors.

In our research we investigated differences in responses toward information and entertainment technologies between U.S. and German consumers. The two cultures were chosen because they represent comparable income levels and penetration of common consumer technologies (Mundorf, Dholakia, Dholakia, & Westin, in press), but different lifestyles. Some of the contextual patterns observed in contemporary Germany are attributable to the unification of East and West Germany and the resultant merger of two very different economic and infrastructure systems.

Content, Form, and Interface

In the previous section we considered various user factors that can impact the adoption and use of information technologies in the home. In this section, we consider features of the technologies themselves. Technologies in the home are adopted and used because, as perceived by the user, they provide some value. Within the realm of information technologies, this value can be supplied through the provision of information, functionality, or entertainment to the user. In considering the traditional information technologies now found in the home—television, telephone, personal computer—examples come immediately to mind. In considering the imminent combining of these technologies into a single device, the examples are less immediate, but still obvious.

Consider a technology, as just noted, that provides the combined utilities presently derived from the telephone, the television, and the personal computer. In this screen-based information technology product, value in the form of information is provided through current, on-demand weather forecasts, sports scores, stock quotes, and the like. Value in the form of functionality is derived through the ability to pay bills by invoking electronic fund transfer transactions, and through the ability to voice a personal opinion in an electronic town meeting. Value in the form of entertainment is provided through the video game with an opponent who is either virtual or lives halfway around the globe.

Moderating each of these three facets of the information technology is the degree of interactivity. In any category, the degree of interactivity can conceivably range from noninteractive to highly interactive.

Interactivity impacts user involvement in an obvious way. The impact on user preference is not so clear. Reports condemning today's youth as a generation of "TV-addicted couch potatoes" would lead us to believe that noninteractive, passive entertainment is the key to garnering mass acceptance and adoption. It is this same group of individuals, however, which is observed to spend hours at a time playing demanding, interactive video games with passion.

In understanding the demand for, and adoption of, an information technology it is imperative that we understand these content and interface factors. What is the demand for information-related services as opposed to functional services in an information technology offering? At what point does the complexity of a service begin to hinder its perceived usefulness? Does interactivity tend to attenuate or increase the acceptance of the entertainment aspects of a product or service? The experimental study described later in this chapter represents an attempt to address some of these issues with respect to screen-based information services. Of particular concern in this study is a dimension of information technology that is often overlooked by providers. This is the entertainment dimension.

A glance at television and radio audience figure reveals that entertaining materials take up by far the greatest share of programming watched or listened to. By analogy, one might expect that the entertainment quality of new information technology content will be a critical determinant of user acceptance (Landler, 1994; MacFarland, 1990).

Hedonic Valence

User response to the entertainment qualities of media content has been found to be a function of physiological arousal in combination with cognitive dispositions (Zillmann, 1980). One key factor influencing the cognitive disposition is the pleasantness or hedonic valence of a stimulus. Research on television and film has shown that this "hedonic valence" is an important factor mediating viewer response (cf. Zillmann, 1988). By implication, it should also affect the way users respond to other types of information media, such as those considered herein.

Mass media research has conceptualized negative and positive hedonic valence. Negative hedonic valence has been operationalized as "gory" scenes (Zillmann, Bryant, Comisky, & Medoff, 1981), positive hedonic valence as pleasant nature scenes or—in some studies—erotic depictions (Baron, 1974). In considering screen-based information services, the concept of negative hedonic valence may not be meaningful. Although violence is becoming an issue for video games, one would not expect to find content that is comparable to horror movies or shocking TV documentaries in services that are mainly text-based. Consequently, a more appropriate dichotomy would seem to be high (positive) versus low (neutral) hedonic valence.

Perceived hedonic content would be expected to be greater for information content if it provides greater sensory enhancement. At present, a screen-based information service allows for enhancement of two senses—visual and auditory. Although enhancements appealing to the senses of touch and smell are conceivable in the future, they are uncommon at the present time.

Color. The most basic type of visual enhancement is color. Research on color has shown numerous psychological and physiological effects (cf. Kuller, 1981; Mikellides, 1990). In the past, for photography, film, and television, the introduction of color was heralded as one of the critical innovations that led to far greater enjoyment (Alber, 1985), and for which users—in the case of television—were initially willing to pay a multiple of the price of the monochrome version. Also, many computer users consider color a desirable feature and are willing to pay a considerable surcharge for a color system or for the upgrade of a monochrome system.

Graphics. A more complex type of visual enhancement is provided by computer graphics. The past decade has experienced an enormous diffusion of computer graphics into all visual media, from "MTV" to *USA Today.* Also business, academia, and even politics bear witness to a proliferation of computer-generated graphics. In addition, the most successful home entertainment game systems in recent years, Sega and Nintendo, are entirely based on animated computer graphics. Presumably, then, the inclusion of graphics in an information service would lead to greater enjoyment derived from the system, and greater user acceptance.

Music. Plenty of research underscore the pervasive nature of music, its capability to affect listener responses, and its propensity for mood management. Many Americans spend several hours a day listening to radio—mostly music. Television programs contain a considerable amount of music, and even shopping malls use music to provide a pleasant backdrop. The use of music would thus be expected to impact user acceptance of a new information technology in a positive manner.

RESULTS: THE IMPACT OF AGE, GENDER, AND CULTURE

In order to assess computer responses to information technologies, a survey of college-age and older respondents was conducted.

A Familiarity and Lifestyle Survey was initially administered to U.S. college students and nonstudents representing all groups. It included the following composite measures: technology seeking, familiarity with technology, and attitudes toward technology in society (Mundorf et al., in press).

Gender Effects

The findings for the U.S. college student sample indicate that men are more familiar with and interested in the high-tech and entertainment features of communication technology (stereo, multiple-window TV, programmable remote control), whereas women are more familiar with technologies that enhance or control interpersonal communication (answering machine, cordless phone). Males show greater leaning toward technologies that have only recently been introduced to the mass market and those that emphasize user control over mediated/entertainment content (CD-interactive, multiple-window TV). These findings corroborate gender differences found in earlier research. One source of male bias is the information or program content of the new technologies. If information content is easily available and appealing to both genders, male bias toward a new information technology should be mitigated.

Males showed greater intention to use a number of technologies. These technologies comprised the categories of passive (stereo and window TV, digital audiotape), and active entertainment (video games), as well as pragmatic utilities.

Cross-Cultural Effects

The survey was translated and administered to the corresponding college-age and older adults in Germany (Mundorf et al., in press). The data supports the view that U.S. consumers are more familiar with most information technologies than German consumers. This difference is especially pronounced for U.S. college students.

Age and Gender. Overall, males tend to be more familiar with technologies than females, younger subjects display greater familiarity than older ones, and Americans are more familiar with technologies than Germans. The data confirms that across cultures, men are more technologically oriented than women. This gender effect was not found for phone technologies. Probably women feel a relatively greater level of comfort with these compared to other technologies, which results in familiarity levels similar to that of males.

An age effect materializes for the composite measure of computing. Apparently, those of college age are much more familiar with computers and related technologies than older groups—regardless of country or gender.

The only clear finding for new technologies was that Americans show greater familiarity compared to Germans. Because very few respondents are aware of some technologies (e.g., CD-interactive), other effects were weak or insignificant.

There is no significant difference in familiarity between German students and older German respondents, whereas U.S. students reveal a greater level of familiarity with technologies than both older Americans and the two German groups. Also, for the more common types of technologies (TV- and phone-based), older Americans show greater familiarity than the two German groups.

Technology Seeking. This trait was measured as a composite score on a battery of forced-choice statements. Americans displayed a greater degree of technology seeking compared to German respondents. Overall, students showed greater technology seeking than their older counterparts. No significant gender differences emerged. But, older German males displayed a more positive attitude toward technology compared to their college-age counterparts.

The hypothesis that across cultures, college-age consumers are more technology oriented than older consumers was not supported for the German sample. German college students are far more conservative regarding the acceptance of technologies than their U.S. counterparts. They are roughly at the same level as older Germans.

Technology in Society. Sixteen statements were developed to investigate attitudes toward technology. Americans showed significantly stronger agreement with three general statements pertaining to the role of technology in life. They concede that it improves our quality of life and gives us control over nature. However, somewhat surprisingly, they also caution that, "With each technological innovation, we lose some essential human quality."

Germans revealed significantly greater concern than Americans did with specific technological advances, notably in the areas of the environment, medicine, human life, and society. However, they also agreed more strongly than the U.S. respondents that "Compared to all past centuries, we live in the best of times."

Older German women disagreed more strongly with the statement "Technology gives us control over nature" than the three other groups. In addition, German males were less concerned than all other groups with the statement "Technology is making our life impersonal."

Germans tend to be more skeptical of the role of technology in society than Americans, especially as far as specific societal areas are concerned (e.g., medicine and the environment). Americans appear to have a more global concern about the loss of essential human qualities due to increasing reliance on technology.

Table 9.1 summarizes the results presented in this section.

TABLE 9.1
The Impact of Age, Gender, and Culture on Technology Attitudes and Familiarity

Variable	*Key Findings*
Age	Greater familiarity with computers for younger group
	Greatest familiarity with technology for younger Americans
	Older Americans show greater familiarity than both German groups
	Older Germans show more positive attitude than younger Germans toward technology
Gender	Men are more familiar with technology than women
	Men are more familiar with recent technologies
	No difference for phone-based technologies
	Men show preference for entertainment technologies
Culture	Americans show greater familiarity with technology than Germans
	Age difference in technology seeking only for Americans
	German college students are more conservative than U.S. college students
	Older German males show greater technology seeking than younger German males
	Germans show greater concern over technological advances
	Germans show greater concern regarding the impact on the environment and medicine
	Americans show concern regarding loss of "essential human qualities"

RESULTS: CONTENT, FORM, AND INTERFACE

In an experiment, subjects believed that they were evaluating a prototype of a new commercial information service. In reality, the respondents were exposed to a software simulation of an interactive videotex system. In this study, hedonic features of the interactive information service (presence or absence of color, graphics, and music) were factorially manipulated. At the end of the 15-minute exposure period, subjects' reaction to the system was measured through an exit questionnaire.

Information Service Simulation

Simulation Mechanism The various experimental treatments were administered through the use of *INFOSERVE*—a custom software system that emulates a complete commercial videotex product. This PC-based simulation provides the experience of using a full, working information service similar to the commercially available Prodigy system. Details of the system architecture can be found in Westin, Mundorf, and Dholakia (1993).

With the *INFOSERVE* simulation mechanism, users are presented with a system of menu-based topic and subtopic choices. Choosing a topic leads to a screen containing a menu of subtopic options. Selection of a subtopic item results in the display of an information screen (such as local cinema features) or a screen requiring user actions (such as e-mail or banking transactions). The system is programmed to allow experimental subjects to freely explore all aspects of the information service at their own pace for an undisclosed period of time. At the end of the fixed exposure period, in this study 15 minutes, the software interrupts their exploration and presents final instructions.

Product Offerings. In this set of experiments, subjects were able to select 10 different main topic categories with 5 subcategories within each main category. Main topic categories included such choices as single scene, job market, shopping, and travel information. In order to increase the realism of this study, time-dependent information content, for example weather statistics, theater venues, and sports scores, was updated before the experimental sessions.

Note that the system offerings provided utility in terms of information (e.g., weather report), functionality (e.g., electronic banking), and entertainment (e.g., humor). Note also that the subtopic choices ranged in degree of interactivity from one-way communication (e.g., sports scores) to moderate user involvement (e.g., in the humor category users could contribute their own jokes) to complete interactive communication (e.g., e-mail). Also, in order to minimize gender bias, care was taken to select some topics and subtopics that might appeal more to females and some with greater male appeal.

Hedonic Factors. As indicated earlier, effects of hedonic aspects of an information service were investigated by factorially varying the presence of color, graphics, and music. In the "graphics-present" conditions, a 2-second graphic image relevant to a subtopic appeared just after that subtopic was selected and before the succeeding information screen appeared. In the "music-present" conditions, a thematically appropriate computerized tune was provided as a 10-second audio overlay to an information screen. For example, the weather report was accompanied by the tune "Rain Drops Keep Falling on My Head." In the "color-absent" condition, the otherwise multicolored screens were presented in a monochrome format.

Measurement and Results

Experimental measurement included both online and offline measures. Offline measurement was in the form of an exit questionnaire that tapped various dimensions of users' reactions and impressions of the system.

Factor analysis resulted in the emergence of three clearly defined factors: hedonic response, intention to use, and difficulty. Composite factor scores were then analyzed by means of analyses of variance. The independent variables used in this analyses were color, graphics, music (present/absent) and gender (male/female). A detailed description of the measurement and analysis can be found in Mundorf, Westin, & Dholakia (1993).

Enjoyment and Intention to Use. Hedonic response measures the degree to which using the service is pleasant and enjoyable. Music increased the enjoyment of the service significantly. This effect was even stronger when music was combined with color. Females enjoyed the monochrome version more than the color version.

Another measure was created that combined respondents' intention to use the service at home, school, or work with their willingness to pay for the service. Women showed considerably stronger intention to use than men did. For both genders, the combination of color and music enhanced the intention to use, whereas music or color alone failed to enhance it.

Topic Choices and Preferences. In addition to the offline data collection and analysis just described, a number of online measurements was also taken. To attain an entirely unobtrusive record of subjects' choices, the software was designed to record and time stamp each selection made by each individual subject. This generates a log of the pattern and duration of all user actions and choices.

This subject activity data that is captured by the *INFOSERVE* system provides a variety of user preference metrics that are impractical with traditional techniques. One measure of user preference is aggregate time spent with a topic. Aggregate time is, in effect, the user's "investment" in the topic area.

Among the 10 main topics, subjects spent the most time with humor (224.9 seconds, on average), followed by singles (89.6 seconds) and sports (71.9 seconds). Males spent more than twice as much time with sports scores (98.9 seconds) as did females (45.1 seconds). Shopping revealed the reverse pattern; females spent 50.3 seconds, whereas males invested only 33.8 seconds.

Topic preferences were also measured by examining subjects' first choice of topics. Sports was chosen first by 25% of the subjects, followed by humor at 19.4%. Most subjects preferred to delay exploration of the more functional topics, such as banking (1.9%) and e-mail (1.9%).

A broad range of behavior is illustrated by the time taken to select the initial topic choice from the main menu. The range is from 10 to 61 seconds. Females appeared more conscientious, taking 24.5 seconds on average, compared to 19.7 for males.

These results begin to shed some light on the effects of content, form, and interface factors with respect to user acceptance and preferences. Although the study described here focuses on screen-based information services, the research approach should be applicable to a variety of information technology products.

The results discussed in this section are summarized in Table 9.2.

TABLE 9.2
The Impact of Content, Form, and Interface of Information Services (College-Age Sample)

Variable	*Key Findings*
Content	Humor, singles, and sports are the most popular topics
	Males show greater interest in sports
	Females show greater interest in shopping
	Females choose topics more slowly
Form	Music increases enjoyment
	Music and color combined increase enjoyment and intention to use
	Females prefer the monochrome version
	No effect for graphics
Interface	Females show greater intention to use simulated information service
	Younger people are more familiar with PC-type interface
	Males enjoy technical gadgets
	Americans are more accepting of technologies in society and everyday life

CONCLUDING REMARKS

The survey studies and simulation experiments discussed here are part of an ongoing program of research investigating the nature and impact of information technologies in the home.[1]

The surveys revealed generally greater technology acceptance by men compared to women, by older compared to younger users, and by Americans compared to Germans. In light of these patterns, it may be desirable to design and market technologies in such a way as to increase acceptance by women and older users. Also, some cultural differences in attitudes toward technologies might be addressed.

Issues of Gender and Age

It appears that fundamental differences remain in the conceptual frameworks that men and women apply toward communication and computer technologies. Women seem to emphasize the communication component, whereas men are focused on the functional aspects of technologies. Women's focus on cordless phones and answering machines indicates that in using technology women are geared toward improving or controlling interpersonal communication. Men, on the other hand, appear to be more interested in entertainment that can be manipulated through technical enhancements.

The experimental findings indicate a reversal of the traditional gender bias toward technologies. Women reported that they enjoyed the prototype system to a higher degree than men. They found it easier to use, and perceived higher practical usefulness in the face of time constraints and the need to make decisions concerning important purchases. Women also demonstrated a greater perceived value of the information service by reporting that they would be willing to pay a significantly higher monthly fee. This preference by women was robust; it surfaced despite the presence of some clearly male-oriented content choices in the system. This suggests that the ease of use and the availability of varied content appealing to both genders in this technology was a critical determinant of female acceptance. One might infer that corresponding changes could also enhance female acceptance of other technologies.

Careful modification of form, user interface and information content may also serve to mitigate the age bias in the use of technologies. Just as the mainstream media have recently begun to cater to an older segment of

---

[1]The research described herein is sponsored by the Research Institute for Telecommunications and Information Marketing (RITIM) at the University of Rhode Island.

consumers, so, too, will the providers of information technologies face this challenge in the future. There is a great need for research in this arena.

Issues of Culture and Nationality

From the standpoint of global marketing it is critical to cater to the tastes and preferences of consumers in cultures other than the United States. The research discussed here illustrates several areas where the perception of technology, as well as its software and interface characteristics, can be modified so as to be more appealing to an international audience.

As would be presumed from the differences in the information technology contexts in Germany and the United States, the U.S. sample was more familiar with new information technologies, especially emerging technologies. A key factor may be that new technologies are becoming available to U.S. business and home markets simultaneously, but there is also a gradual erasing of boundaries between work and home.

As expected, Germans are skeptical about the benefits of new technologies. At the same time, however, they recognize the benefits that technology has brought about.

At a deeper level, Americans may be disturbed about the fact that their simpler past is being rapidly replaced by hyper-technological culture and that human relationships are being endangered. This may be especially the case since technological change comes on top of the turbulence brought about by the rapid pace of social change (job loss, family breakups, organizational downsizing, etc.) in the United States. This may be why more Americans than Germans agreed with the statement "with each technological innovation, we lose some human quality."

The skepticism toward new technology, coupled with relative unfamiliarity with the technologies, means that the German household market is likely to be less welcoming to new information products and services than is the U.S. market. U.S. firms are entering into partnership with established European firms to gain the acceptance of their offerings in German markets. Even after such alliances are formed, the general skepticism toward technology is likely to remain an obstacle and specific strategies may have to be formulated to counter or alleviate such skepticism. Nixdorf, for instance, is trying to create a niche in the PC market by advertising its environmental responsibility. New technology markets will also have to be more respectful of the boundaries between home and work than they are in the United States.

Even in the United States, technology providers will have to address the deeper level of anxiety about technology which is perceived to be destroying human qualities. Marketers will have to be sensitive to the human relationships that people value. This is already being attempted, for example, in the marketing of telecommunication services by companies like AT&T which advertise the joys of connecting across long distances. Advertising and promotion, however, are not enough to overcome the fundamental anxiety about the perceived dehumanizing quality of technology. Ultimately, it is the design and distribution of technological products and services that helps preserve and even build human relationships.

REFERENCES

Alber, A. F. (1985). *Videotex/Teletext.* New York: McGraw-Hill.

Antonoff, M. (1989). The Prodigy promise. *Personal Computing, 13,* 66–78.

Arch, E. C., & Cummins, D. E. (1989). Structured and unstructured exposure to computers: Sex differences in attitude and use among college students. *Sex Roles, 20,* 245–254.

Armstrong, L., Yang, D. J., & Cuneo, A. (1994, February 28). The learning revolution. *Business Week,* pp. 80–88.

Baron, R. A. (1974). The aggression-inhibiting influence of heightened sexual arousal. *Journal of Experimental Social Psychology, 10,* 23–33.

Beville, H. M. (1988). *Audience ratings.* Hillsdale, NJ: Lawrence Erlbaum Associates.

Brownell, W. E., & Mundorf, N. (1990). *Student attitudes towards computers and the willingness to communicate on-line.* Unpublished manuscript.

Carey, J. (1991, October 23). *Consumer adoption of new communication technologies.* Lecture presented at The University of Rhode Island, Kingston.

Carter, B. (1994, January 3). 500-channel TV: Around a distant corner. *The New York Times,* p. C11.

Day, K. D. (1980). *The effect of music differing in excitatory potential and hedonic valence on provoked aggression.* Unpublished doctoral dissertation, Indiana University, Bloomington.

Ferguson, D. A. (1991, November). *Gender differences in the use of remote control devices.* Paper presented at the annual convention of the Speech Communication Association, Atlanta, GA.

I can't work this ?[@]!!@[|] thing. (1991, April 29). *Business Week,* pp. 58–66.

Kerschner, P. A., & Chelsvig, K. A. (1981). *The aged user and technology.* Paper presented at the Conference on Communications Technology and the Elderly: Issues and Forecasts, Cleveland, OH.

Kiesler, S., Sproull, L., & Eccles, J. S. (1985). Pool halls, chips, and war games: Women in the culture of computing. *Psychology of Women Quarterly, 9,* 451–462.

Kubey, R., & Larson, R. (1990). The use and experience of the new video media among children and young adolescents. *Communication Research, 17,* 107–130.

Kuller, R. (1981). *Non visual effects of light and color, annotated bibliography.* Stockholm: Swedish Council for Building Research (Document D15).

Landler, M. (1994, March 14). Are we having fun yet? Maybe too much? *Business Week,* p. 66.

Lockheed, M. E. (1985). Women, girls, and computers: A first look at the evidence. *Sex Roles, 13,* 115–122.

Lohr, S. (1994, January 3). The road from technology to market place. *The New York Times,* p. C11.

MacFarland, D. T. (1990). *Contemporary radio programming strategies.* Hillsdale, NJ: Lawrence Erlbaum Associates.

Meyer, S., & Schulze, E. (1993). *Projektergebnisse: Technikfolgen fuer familien* [Project results: Effects of technology for families]. Duesseldorf: VDI-Technologiezentrum.

Meyer, S., & Schulze, E. (1994). *Alles automatisch: Technikfolgen fuer familien* [Everything's automatic: Effects of technology for families]. Berlin: Edition Sigma.

Mikellides, B. (1990). Color and physiological arousal. *The Journal of Architectural and Planning Research, 7*(1), 13-20.

Mundorf, N., Dholakia, R. R., Dholakia, N., & Westin, S. (in press). Orientations towards Germany and the U.S.: Implications for marketing and public policy. *Journal of International Consumer Marketing.*

Mundorf, N., Meyer, S., Schulze, E., & Zoche, P. (1994). Families, information technologies, and the quality of life: German research findings. *Telematics and Informatics, 11*(2), 137-146.

Mundorf, N., Westin, S., & Dholakia, N. (1993). Effects of hedonic components and user's gender on the acceptance of screen-based information services. *Behavior & Information Technology, 12*(5), 293-303.

Nussbaum, J. F., Thompson, T., & Robinson, J. D. (1989). *Communication and aging.* New York: Harper & Row.

Rogers, E. M. (1986). *Communication technology.* New York: The Free Press.

Rooney, A. (1991, May 12). *60 Minutes.* New York: CBS Television.

Salvaggio, J. L., & Bryant, J. (Eds.). (1989). *Media use in the information age: Emerging patterns of adoption and consumer use.* Hillsdale, NJ: Lawrence Erlbaum Associates.

Schlossberg, H. (1991). Interactive TV forges ahead. *Marketing News, 25,* 1ff.

Sisodia, R. (1991). *Singapore—Toward an "Intelligent Island"* (Working Paper No. 91-04). Kingston: Research Institute for Telecommunications and Information Marketing, The University of Rhode Island.

Strom, S. (1994, January 3). Testing the high hopes for TV shopping. *The New York Times,* p. C12.

Venkatesh, A., & Vitalari, N. P. (1986). Computing technology for the home: Product strategies for the next generation. *Journal of Product Innovation and Management, 3,* 171-186.

Venkatesh, A., & Vitalari, N. P. (1987). A post-adoption analysis of computing in the home. *Journal of Economic Psychology,* 161-180.

Westin, S., Mundorf, N., & Dholakia, N. (1993). Exploring the use of computer-mediated communication: A simulation approach. *Telematics and Informatics, 10*(2), 89-102.

Zeithaml, V. A., & Gilly, M. C. (1987). Characteristics affecting the acceptance of retailing technologies: A comparison of elderly and nonelderly consumers. *Journal of Reading,* 49-68.

Zillmann, D. (1980). Anatomy of suspense. In P. Tannenbaum (Ed.), *The entertainment functions of television* (pp. 133-164). Hillsdale, NJ: Lawrence Erlbaum Associates.

Zillmann, D. (1988). Mood management: Using entertainment to full advantage. In L. Donohew, H. E., Sypher, & E. T. Higgins (Eds.), *Communication, social cognition, and affect* (pp. 147-169). Hillsdale, NJ: Lawrence Erlbaum Associates.

Zillman, D., Bryant, J., Comisky, P. W., & Medoff, N. J. (1981). Excitation and hedonic valence in the effect of erotica on motivated intermale aggression. *European Journal of Social Psychology, 11,* 233-252.

Zillmann, D., Weaver, J., Mundorf, N., & Aust, C. (1986). Effects of an opposite-gender companion's affect to horror on distress, delight, and attraction. *Journal of Personality and Social Psychology, 51,* 586-594.

CHAPTER 10
Literacy in the Age of New Information Technologies

A. Fuat Fırat
Arizona State University West

Given the momentous developments in information technologies, the issue of literacy requires a rethinking. Such rethinking is linked closely to the phenomenon of postmodernism, because postmodernist insights help to provide some of the most enlightening analyses of these developments (see, e.g., Poster, 1990). However, when both the development of new information technologies and the phenomenon of postmodernism are involved, one encounters highly complex and varied issues. Consequently, it is impossible to cover the total ground. An attempt is made to address those circumstances that especially pertain to literacy. To provide the necessary context for such a discussion, first we should consider the characteristics of modernism as they relate to the topic at hand.

MODERNISM

As is widely recognized by many philosophers of our time, modernism was centered around a project (see, e.g., Habermas, 1983; Lyotard, 1984; Rorty, 1979). The project of modern society was to liberate the knowing subject (the human being who informed oneself about the reality around her or him and acted upon it, specifically on the objects in the environment, based on this information or knowledge) from the limitation imposed by nature. Modernity strived to improve human life by controlling nature through scientific technologies (Angus, 1989). The knowing subject was a modern subject as opposed to the *being subject* of antiquity who was servile to a higher authority, a supreme being. Modern society was built upon the idea that the human being, the knowing subject, was or could be in control of one's own life rather than live out a fate that was determined beyond his or her own control. With this idea, modernity turned its eyes onto the future, determined to fulfill the promise of a better life for humankind, committed to this project.

Thus, the modern project put the human being into the center of things, as the one who was to benefit from the project, and the one who would realize the project. To be able to have a project with attainable goals, to work toward the realization of reasonable goals, order was needed. Chaos and crisis, after all, are not conducive to achievement of set, recognizable goals. Thus, while having disposed of the idea of a theological, spiritual order, modernity did not eliminate the idea of order. The idea of a spiritual order was replaced by the idea of a material order, one that "truly" existed and constituted humanity's reality.

The Modern Subject

Modern society constructed its ideal of order specifically by separating domains or spheres of activity and purpose. As Habermas (1983), one of the major philosophers of late 20th century, articulated through building upon Max Weber's conception, modern society had to develop norms to uphold the idea of an order. Consequently, a search for normativity was a principal quest for modern society (Steuerman, 1992). Yet, the norms had to support the possibility of the existence of a knowing, reasoned subject, and when the emotional and value based dimensions of being "human" entered the picture, the construction of such a knowing subject was tainted. In the search for normativity, therefore, occurred the separation of three spheres: science, art, and morality. Each sphere, then, could be governed according to its own inner logic (Foster, 1983), without contaminating the others; especially, of course, the sphere of science had to maintain its purity to allow objective knowledge. Thus, to the sphere of science were assigned the norms of knowledge, truth, and reason. The sphere of art was assigned the norms of beauty and aesthetics. The sphere of morality was assigned the norms of justice and normative rightness (ethics). Mixing the spheres and their inner logics would derail the progress of humanity towards its project, the higher purpose of improving human lives.

Modern philosophers and scholars explored, investigated, and established norms for each sphere. Within the sphere of science, norms—for example, in terms of the correct method(ologie)s—were *discovered,* according to the rationale of objective knowledge, based on reason. These norms, then, provided the proper and improper ways and means of acting in each sphere. Consequently, modernity produced many bipolar, oppositional categorizations guiding the proper and the improper, therefore, the superior and inferior ways of action. Rational–emotional, public–private, production–consumption, culture–nature, mind–body, masculine–feminine, active–passive were some of the most important categorizations that enlightened the way to understanding the order that existed and that would guide progress toward the modern project. Very important among these was the separation between the mind and the body.

It was this Cartesian separation—between mind and body—that enabled the realization of the knowing subject (Russell, 1945). The body was immersed in, determined by, and dependent on, therefore, limited and constrained by its nature. Being so immersed in the material conditions of its own existence, the body could not very well take an independent, objective position in discovering the principles of its reality; it was biased by its own nature. Knowing, on the other hand, required objectivity and independence, and the ability to distance one's powers of cognition from what was being investigated, to obtain a detached and disinterested position. How then, could the human individual so immersed in nature—in the material conditions of existence—become a knowing subject with the ability to recognize what were true representations of reality and what were interested and biased representations? Somehow, a separation from the immersion into nature had to be achieved, and this was attained by the separation of the mind and the body. The human mind, although attached to the body, could detach and separate itself from the everyday, profane experiences and existence of the body, thereby being able to attain an objective stance and knowledge.

Modern Literacy

In modern thought, the medium of the mind was (is) considered to be language. Language, claimed to be the major quality that differentiated the human from the beast for the modern scientist, was not part of nature, it was a creation of the human society. Furthermore, the units of language, words, were the instruments through which knowing subjects were able to represent things around themselves, thereby *conceptualizing* their existence and *abstracting* their essences, both feats necessary to separate the objective and the essential from the partial and the apparent. Such conceptual representation of things by signifiers (in modern society, specifically verbal signifiers, that is, words) could be independent of nature, belong to the realm of culture, and enable reasoning through the signified (concepts) that had been extracted and abstracted from the immersion into everyday experiences with nature. Therefore, such a structure of language among the referrents, the signifiers, and the signified could enable the human being to be detached and disinterested, thus, an objectively knowing subject.

The form of language that was permanent, the written word, commanded a special position for at least three reasons. First, it was permanent, thereby preserved for the future; the period toward which modern society had turned its sights. Second, it allowed documentation that could be examined and judged repeatedly, allowing more lasting mental discourse and making it available to larger audiences—especially in early modernity when other technologies of mass communication were not well developed. Third, because of its permanence and availability, it had a materiality to it that the spoken word lacked. Thus, in modern society, literacy, the fundamental form of knowledge, came to be defined as the ability to read and write.

The importance and value of permanence and materiality in modernity were observable in all spheres, not only in science and literacy. In the arts, for example, for any piece of work to be recognized as art, especially as modernity progressed, it needed to have permanence and materiality. Also important, and analogous to the abstraction and conceptualization of things through words that represented them, was the extraction and separation or detachment of the piece of art from everyday life experiences. A piece of art, such as the painting or sculpture, had to be separated from everyday, common usage to command an interest in it as an investment, as *art,* as something that had durability beyond common, functional or utilitarian consumption. Quilting, for example, has gained in stature as art, as opposed to a craft, as more and more quilts were made, not to be used in keeping warm, but to be "exhibited" by hanging on the wall or otherwise.

These impulses to extract, distance, detach, and abstract from common experience are very strong in modern thought and practice. That which is separated and abstracted from daily, common usage and practice, then, is valued as superior, at times, sacred. Immersion into the common experience, on the other hand, is generally considered as banal and profane. In language, that which is valued and superior is that which is most abstract, to be practiced by only the ablest minds, at the greatest distance from the everyday experiences of the body.

Importance of the Visual in Modernity

A combination of circumstances in the construction of modern society resulted in the primacy of vision over the other senses. One, of course, was that reading the written word, the ultimate means of reasoned mental discourse, was a visual experience. Perhaps even more important was the impact of the physical sciences, especially astronomy, which played a pivotal role in the liberation of reason from religious repression (Rorty, 1979; Russell, 1945). Much scientific activity was dependent on *observation* through optic instruments, such as the telescope. When common observation that could be replicated was present, "reasonable" people could no longer deny the existence of material phenomena. Then, what was discovered through observation was immortalized through written words, numbers, and symbols, the substance of reasoned explanations and theories. The mind (reason), the word (specifically, the written word), and sight (the visual) seemed to find a happy and sacred togetherness or integrity in a modern world. Yet in this togetherness lay seeds of paradox that became more apparent with further development of modern technologies.

In the development of modern society, the privileging of the visual is evident in many instances (Levin, 1993). In modern art, for example, the visual arts, such as painting and sculpture, have a special place; they took center stage and received special recognition. The privileging of the visual is also evident in the centrality of the voyeuristic gaze, its eminence in verification (observation) of facts that are the building blocks of scientific knowledge. Being so privileged, when modern technologies of the visual—photography, cinema, and television—appeared, they enthralled the public. They were fulfilling (satisfying) to the sense that was most revered and, therefore, had become most acute. In many cases, they enabled "direct" experience of things that were otherwise too distant and unreachable. They were, indeed, "the next best thing to being there."

Paradoxically, the visual, that which had mutually reinforced and was privileged by the modern status of the written word, now eclipsed it. As modernity progressed, new forms and technologies of visual experience were developed; specifically, photography, cinema, and television. They were mediating forms and technologies. That is, they allowed the observer to "see" events and environments without actually being present at the "scene." They separated the visual experience from direct, unmediated presence earlier required to be a visual witness. In order to "see" the Grand Canyon, or an African Safari, one did not actually have to be there. In providing such detachment, and the ability to see through the "objective" eye—by "catching" the moment for all to see, enjoy, and investigate from the same perspective—these visual technologies gained higher status as art forms as well. With such success, the visual experience also distanced itself from its original partner, the written word, and became its rival. Now people could watch events on film, on television, through photography, and have a visual experience of what happened, get a "truer" and richer account than by reading about them. In many ways, the visual was faster, more "real," more exciting and "spectacular."

Yet, at the same time, the visual technology was closer to unmediated immersion than reading/writing in the sense that it was a "direct" registration of a *sensual* experience. From a modernist way of thinking, it lacked the mediation of language, words, and symbols that allowed conceptualization and reasoning. Therefore, the technologies of the visual were inferior and not part of literacy or of true knowledge. Modern thinkers assigned these technologies the lesser role of entertainment, and they argued that these technologies distracted their audiences from the intellectual pursuits. This may have been true, not because of the nature and potentials of the technologies but because of the cultural uses they were put to, as I discuss later.

Modern thought reacted to the allure and excitement of the visual technologies, some of which also integrated sound, with its usual impulse: it created an oppositional and polar relationship between entertainment and education (learning/knowledge). Cinema and television were media of entertainment, not of education or true knowledge, and their allure, their ability to seduce, was (is) treated with some contempt by the modern literate. According to the modern literate, and to many modern educators, entertainment of the sort provided by television and cinema was distracting the youth, and even the adults, from the more serious and important issues of the day. Furthermore, these technologies of entertainment are blamed for causing a decrease in reading activities and contributing to the crisis in literacy. According to some recent surveys, for example, over half of the U.S. residents with college degrees cannot consistently figure out how much change they should get back from the lunch bill or how to read a complicated bus schedule ("Most College Graduates," 1993). As such, the visual and the written word are now considered to be rivals instead of partners in the project of progress.

As exemplified well by the experiences of Leslie Stahl (in Moyers, 1989c), the White House correspondent for CBS network news during the Reagan presidency, the visual overrides the verbal even in the serious business of news, grabbing the attention of the audience away from the linear logic of words. As articulated by Michael Deaver (in Moyers, 1989c), a Reagan advisor, this circumstance was well recognized and employed by the Reagan White House. Indeed, one of the argued qualities of the (written) word was (is) that it is conducive to a linear, analytical logic, enabling a structure of thought to be built upon an ordered, sequential set of foundations. The visual imagery, on the other hand, is much more conducive to being multilayered and multifaceted. Many images coming from disparate realms with differing principles can be simultaneously reflected and represented, juxtaposed and processed (Ulmer, 1990). The visual can be, thus, spectacular, and the spectacle tends to be more seductive than the logical. Possibly, one reason for its allure is that it is sensual. Modern society, in privileging reason and its form, literacy (defined as writing and reading), seems to have put itself into a bind. When technologies that can address the senses and, therefore, the sensual are developed, the goal of progress solely through reason and modern literacy seems to falter.

POSTMODERNITY AND THE FORM OF LITERACY

If, indeed, the visual and other technologies of the sensual are so alluring and distracting from reason, can the modern project of improving human lives by emancipating the human being from impositions of nature, debilitations of ignorance, and exploitation of each other still be accomplished? Should society ban the technologies of the sensual or regiment them to continue the progress toward the modern project because these sensual communication media create a society of voyeurs and onlookers who can mentally process the plethora of visual images and other multidimensional, multilayered signs less and less, and therefore, lose their ability and interest in participating in the social and political processes that determine their fates more and more (Moyers, 1989a)? Neither banning nor regimentation have very positive implications. Yet, newer technologies that will make sensual communication even more seductive and spectacular are developing. These include integrative technologies that combine the television, the computer and the telephone, and virtual reality technologies that will enable immersion of the human individual into a total sensual experience, providing sight, sound, touch, smell, and taste. Further, the population seems to be ready to be immersed into experience rather than stand aloof and distanced (Sorkin, 1992). How are the modern principles of literacy, education, and knowledge to be sustained under such attack?

Postmodern Insights

Postmodernists have argued that the separations upon which the modern ideas of order and the modern project were constructed were mythical from the very beginning, as any culture is a mythical construction. According to the postmodernists, such separations are not and were not possible at any time except in the imaginary (prevailing ideological narratives) of modern societies. Even if the separations between mind and body, intellect and entertainment were possible at any time, they are no longer possible given the world of new information technologies. We may now realize that any such separation might have been remotely possible only because the means and technologies of communication except the verbal ones were underdeveloped.

Postmodern insights remind us of the multidimensionality of humanity and that reason is only one dimension of being human. Recent experiences in modern society—such as pollution, increasing regional and ethnic wars, growing allocation of resources for military purposes, intensification of poverty and misery in large segments of the world's population, and increased crime and violence in the, so called, affluent societies—may clearly indicate that human lives have been disenchanted by the modernist narratives that forced the unidimensional privileging of a singular form of reason. There is a growing willingness exhibited by members of (post) modern society to immerse themselves into situations rather than take the remote position of an "observer." This is witnessed in the increasing thematization of living environments and the growing success of theme parks (Kroker, 1992; Sorkin, 1992). Furthermore, there seems to be an increasing distrust in the project of modernity that asked modern subjects to commit themselves to the vision of a promising future, to sacrifice for it, and to conform to the "scientific" policies that would take them there. There is a growing feeling that, instead, despite some major medical breakthroughs, modern society has delivered much mechanization, misery, and ecological decline. Consequently, postmodernists detect an erosion of commitment to any single narrative or project, leading to a fragmentation of purposes, orientations and, in general, ways and forms of existence. Also detected is a merging of domains, spheres, genres that were so deliberately separated in modern society. It is no longer possible to clearly distinguish architecture from landscaping or art, art from literature, theater from dance, or even politics from fiction—as evident in the Murphy Brown versus Vice President Dan Quayle episodes. Maybe most important is the growing recognition that spheres of science, art and morality are inseparable, and the acceptance that they ought not be separated, that they should inform each other. This is analogous to what Ulmer (1990) urged:

> We are not going to make the same mistake that was made in the Renaissance, when the practices developed to exploit the book required the suppression of pattern (magical) thinking that had survived for centuries alongside the alphabetic culture. In the same way, oral culture was excluded from the cognitive hierarchy. Those interested in electronic technology are not going to make that same mistake. We are allies, not enemies, of the book, because the two kinds of thinking, and the two kinds of technologies, are not in competition. They are complementary. (p. 18)

Postmodernists argue that privileging of any one sphere, technology, or form of knowledge over others, as science was privileged in modernity, leads to tyrannical systems of social narratives and oppression (Lyotard, 1992). Examples often provided for this argument include the Holocaust as well as the developments in weapons of destruction largely imposed by the single-minded narratives of superiority and the Cold War. The feminist literature abounds with examples of how patriarchal forms of knowledge, so dominant in modern society, have resulted in experiences of oppression (see, e.g., Butler, 1990; Butler & Scott, 1992; Harding, 1986; Nicholson, 1990).

The Need for Postmodern Literacy

What seems to be apparent is that given the disillusionment with modernity, the attraction to immersion rather than to detachment on the part of the (post)modern consumers, and the emergence of new information technologies, it is not possible to function simply with the modern form of literacy. The individual of the contemporary (post)modern world is bombarded with sign(al)s, data and information coming from all sorts of different media and in all kinds of different forms. Furthermore, they do not come with a linear logic, in a manner that is *ordered* in the modern sense. Even throughout modern society, the communicative practice rarely followed modern reason. Although modern discourse insisted on the necessity of using reason, modern communicative practices, especially those that had the active purpose of persuasion, such as political and commercial advertising and propaganda, kept trying to circumvent reason by creating means of reaching and affecting emotions directly to influence human behavior. Advertisers and political campaign advisors understood the importance and impact of emotions and values. As Richard Wirthlin, director of consumer research for Ronald Reagan's presidential campaign in 1980, and winner of the Advertiser of the Year Award for his efforts in this capacity, clearly articulated (in Moyers, 1989b), communicating, that is getting across ideas and meanings, thereby exciting thought and, in the end, eliciting intellectual conclusions and understanding—along with emotional responses—cannot be achieved singularly through reason and logic, but requires a fuller, more complete set or arsenal of sensibilities and sensations. Or, put in other words, communication that employs the full set is more effective in relaying its message and eliciting the desired effect. Modernists believed that when means other that reason and words were used, the desired effect could not be intellectual but only emotional; only to circumvent reason and move the audience into senseless action. From such a modernist perspective, each time modern education provided the means (tools) of *reading* (making sense of) communicative media, modern technologies produced new media and means to transmit *unreadible* but *deeply felt* messages (sets of signals). The advent of television, the leading medium of contemporary society, and of advertising, which is increasingly becoming the dominant form of public discourse (Ewen, 1988; Miller, 1989), attests to this continual struggle between reason and emotion created by modernist myths about education and (or versus) entertainment.

Advertising is, indeed, that form of public communication and discourse, and even that art form—although admittedly mostly commercial—that seems to command the greatest resources and many creative minds in contemporary society. As such, and also because it is the culmination and coordination of our latest knowledge and technology for communicating persuasively, it has become the model for practically all other forms of communication. Music videos, movies, television, and radio programs, even the newspapers are incorporating, at a fast pace, the different aspects of advertising. Some of these aspects include the continual polling of the audiences (markets), use of varied combinations of message structure, content, and format that are found to be influential in capturing attention and in persuading, employment of symbols, icons, and other cultural signifiers that seem to be effective in transmitting meanings and feelings, and, in general, the pace and style of advertising. Perhaps most meaningful for our discussions is the increasing use of different forms of sign(al)s, nonverbal as well as verbal. Every medium, televisual, print or audial, is increasing the usage of all possible sign(al)s, and, of course, there is the advent of the integration of different media, and, therefore, the *merging of genres* in communication forms. Music videos, for example, are very much advertisements, not only because they continually flash brand names, but because they serve, for many sponsors, as one of the most effective ways to represent the styles (lifestyles, clothing styles, etc.) to be promoted to vast target markets reached by these videos. On the other hand, advertisements incorporate aspects of other communication forms. An example is the recent development of the "serial advertisement" (such as the Taster's Choice television commercials), reminding one of the soap operas or other series on television. With this merging of genres comes the merging of the different media, technologies, and sign(al) systems of the genres. In the end, we are encountering the incessive development of the multidimensionality and multilayeredness of all forms of communication. To try and stand against this tide by arguing for the specialty of the word seems utterly futile and, also, shortsighted.

It is shortsighted, because, as already mentioned, insisting that the word is superior tends to contradict all experience. The advertising medium has been a good example of how people learn and come to conclusions regarding "reality" from what is communicated to them in visual images as much as, if not more so, from what is communicated to them in words (Kilbourne, 1987). Visual role models in advertisements have been more influential than verbal modeling. Another example relates to learning gender significations by children. They learn much from the nonverbal cues they receive about what it means to be feminine and masculine, and no matter how much they may be told otherwise, orally or in written form, their experiences with what they see tend to constitute much of their consciousness about being male or female (Williams, 1986). This is, by no means, limited to children. Adults are just as much influenced by their nonverbal experiences, and this shows repeatedly in the literature on the body. How people feel and think about their bodies tends to be highly influenced by their visual experiences, especially fashion visuals ("Body Digest," 1987; Jacobus, Keller, & Shuttleworth, 1990). When confronted with this evidence, the modernist impulse may be to react by arguing that all such influence is negative in that it distracts from reasoned, logical processing of information, supplanting it, instead, with uncontrollable and impulsive action. Given that the experience of modern society illustrates that separating emotions from judgment based on reason, or expecting that such a separation can be practiced is, indeed, an illusion—even if it did become partially valid in modern society due to the power of the modernist rhetoric that declared the modern narrative of science, reason, the knowing subject and the modern project—it seems useless to keep hoping and arguing for the separation. At each turn, the development of new communication media—because addressing the sensual and the emotional has proven to be effective in guiding human values, imagination, and behavior—has indicated the inseparability of reason and emotion; that the two are joint qualities of being "human." What may be possible is to integrate or merge the joint qualities rather than futilely seek to separate them.

Indeed, postmodernist discourse provides this integration. It is achieved through the recognition that the contemporary subject is not just the knowing subject of modernity, but a *communicating subject.* In postmodernist discourse, communication is not solely an act of relaying or transferring meaning, but an act of *construction.* It is through communicating, to self and to others, that the subject, the human being, constructs meaning, produces interpretation and reason(ing) to have understanding, and ultimately establishes "reality." The communicating subject is not, therefore, active only in the sense of acting upon his or her environment to investigate and discover his or her material conditions of existence in order to control nature and its effects. The communicating subject constructs his or her social environment, and to a large extent, even his or her immediate material environment. This process of construction is generally termed *hyperreality* in postmodernist discourse (Baudrillard, 1983; Eco, 1986). The concept indicates that hype created by and through effective, powerful communication shapes the preferences, choices, and behaviors of a community in such a way that it becomes the reality of that community (Fırat & Venkatesh, 1993). Modern society, through its industrial strength and the breakthroughs in its communication technologies has (had) become very much a society of the hyperreal. At the same time that modern rhetoric claimed the independence of reality from human agency, preserving the image of the knowing subject as solely one who discovered and represented reality, modern culture, through its industrial technologies, constantly shaped the reality that humanity encountered. Consider, again, the shaping of the material and social human existence through the construction of urban environments—largely based on hypes of promising future comforts and control over nature—specifically, the cities and the suburbs, which greatly formulated the life experiences of the modern citizens. Yet, as modern society shaped human experience and reality to degrees not known before, modernist discourse, especially in the West, continued its insistence that reality was independent of human action. Postmodernist and poststructuralist discourse disputes this narrative. It recognizes the *communicating subject,* one who constructs one's own reality through effective communication that (re)produces hype(s) and/or simulation(s) that is (are) powerful, alluring and seductive. Such communication that constructs reality requires the use of the full spectrum of sign(al)s, that is, words as well as visual images, sounds, and so on, as already discussed. Life in hyperreality necessitates a sensibility that contains reason, yet transcends it. It requires the use of all human qualities, such as, reason, intuition, sensation, and emotion, not excluding but integrating all dimensions, and without diminishing the value of any to privilege a single one.

A modernist is likely to argue that all qualities other than reason will distract from true understanding and judgments regarding the human condition; as if reason and rationality are unique, universal and unerring. Yet, none seems to be the case. Experiences with different cultures (Marsella, Devos, & Hsu, 1985) and with the history of human thought, including the history of science (Feyerabend, 1981, 1987), clearly contradict the claims of uniqueness, universality, and infallibility. Furthermore, the modernist claim that reason is possible only through words because they are the only signifiers that enable cognition, conceptualization, and abstraction, therefore, information and knowledge, is also no longer defensible. As mentioned earlier, semiological findings have indicated the arbitrariness of the linkages of words (as signifiers) to their referrents and their signified (meanings and concepts they represent). Semiologists further have come to recognize, as have postmodernist philosophers (Baudrillard, 1983; Derrida, 1982; Greimas, 1987), that all sign(al)s are signifiers of similar quality as words. Icons, visual images, sounds, and other forms of signifiers all have the ability to conjure up meanings and *inform.* The traditional sense of information, that is, overt, verbal, linear recounting of facts, can no longer be considered the only form of information. After all, a word is just a sign as is the picture of a tree. They can equally elicit cognitive (mental) associations and abstract thoughts. Most of us just do not yet know—although those in advertising, for example, have a much better idea—how to "write" with signs other than words. Otherwise, in reasoning, thinking through, getting meanings across or in interpreting meanings there is really no difference among signs. One can tell a story through words, or visual, sonic images and tunes once these signs are culturally (pragmatically) signified. In the movies, for example, certain tunes will signify impending danger, others happiness, and so on, in ways that words cannot. Each type of sign has its strengths and weaknesses, and under different circumstances different signs will be more successful in provoking meanings, even the sublime (Lacoue-Labarthe, 1986; Lyotard, 1986). Therefore, instead of privileging one or another, the communicating subject can (should, ought to) use all to their highest potential.

Consequently, the modern emphasis on reading and writing, and on the supremacy of the linear logic of reason, to the exclusion of all other forms of discourse in the discovery of truth and knowledge, perception and understanding, in fact, may have hindered the development of truly literate communities. Postmodernists further argue, as mentioned, that such emphasis has disenchanted the experience of being human; it has imposed a stressful human existence by trying to pull apart and separate integral qualities of being "human." The opposing of literacy to entertainment has clearly contributed to the increasing numbers of people who lack even the reading and writing skills because such activity seems so boring and redundant, given the allure, proliferation, and excitement of other media, such as television, movies, and electronic games. The modern insistence on separating these media and forms of communication, then casting them as opponents, instead of integrating them, has not helped, it seems, in creating a literate population.

With the advent of new information technologies based on the computer and its integration with television, virtual reality, and the like, holding on to the modern standard of literacy will further retard the ability of large segments of the population to participate in the construction of the visions and the imaginaries, thereby, the realities of their societies. Those who cannot construct powerful significations and representations of life experiences and their purposes cannot effectively communicate their preferences and choices. Instead, they become a perpetual audience to the representations and significations that are constructed by others who are truly literate, finally having to conform to others' visions and realities. And, those who are not literate in "reading" the multilayered, multifaceted, multimedia signs of communication now increasingly possible, managable and malleable, have no chance in integrating reason into their existence. Given the new information technologies, therefore, a new form and definition of literacy is required to achieve the reintegration, and thereby, reenchantment of body and mind, emotion and reason. Recognition of this new form of literacy is especially needed for the achievement of democratic participation.

Postmodern Literacy

The necessity of a redefinition and understanding of literacy is illustrated by the fact that recently the meaning of text has also changed to include all networks of (informing) signs. Text is no longer just the written material or words in a book but the complete semiotic environment of the human individual. Again, the reason for this redefinition is the recognition that information is received in and through media other than the word, and that the human subject has been influenced and controlled by signs of all sorts.

We find a similar transformation in the meaning of the term language. Body (e)motions, gestures, ornamentation and the like have been increasingly considered as linguistic terms subject to linguistic principles, communicating meanings, intentional or unintentional, just as spoken or written language do. With the extension of text to all signs and/or symbols so is language extended, as is linguistics into semiotics or semiology. Then, of course, there is the advent of hypertext and/or hypermedia (Bolter, 1991; Landow, 1992). Not only is text possible through symbols other than words, but it can be (re)presented in multiple, nonsequential, *intersequential,* overlapping and overlaying sets or networks of sign(al)s that can be seen, heard, tasted, smelled or touched. In a world of such hypermedia, literacy requires an enlarged form of "reading"; that of *sensing*—receiving and processing information in all sensual forms and making sense of it. Concurrently, complete literacy necessitates the ability to *construct* multimedia (hyper)texts—the ability to manipulate, structure, coordinate and, thus, signify all kinds of signs (verbal, sonic, tactile, visual, odorous, etc.).

To argue against this new literacy—we may call it postmodern literacy or *(multi)signefficacy*—would be to insist that signs or symbols (signifiers) other than words (oral or written) do not communicate information or valid information. The fallacy of such insistence is more apparent today than it may have been before. As humans, we have always been informed by signs other than words; by pictures, music, heat, types of smells, color, taste, and so forth. Modern thought probably did realize this, but reduced all to the idea that unless translated into words to be cognitively processed, such information was not valid, therefore, useless. The idea was that *mental* conceptualization required *verbal* language. This was the instrument of the mind and reason because, as already discussed, words extracted and abstracted from immersed experience thereby allowing the detached, independent representation of reality.

Yet now we know, after decades of experience and experimentation with technologies of visual imagery, that meanings, mental images and information are transmitted in nonverbal signs. Advertisers are aware, for example, that visual images, singularly as well as in collages and other juxtaposed forms, do "speak" and leave those exposed to such communication with impressions and information that contribute to construction of knowledge about the product or brand advertised. Thus, the old idea of informative advertising—as opposed to non-informative advertising full of images, jingles, emotional hype, and so on—is no longer valid; if it ever was. Consumers are informed by and through many forms of signs: the attitude of the person using the product in the advertisement, the genre of music in the background, the fleeting colorful collages of activity, and so forth. Each sign indeed speaks and informs the consumer, leaving him or her with a holistic impression as to the nature and meaning(s) of what is advertised.

Clearly, as in the case of information that is transmitted through words, the consumer may find the knowledge or impression communicated to be the same or different from the experience he or she has with the product.

The modern assumption has been that information transmitted through words is possible to validate, given the further assumption that words have clear and absolute, unique meanings, whereas other forms of signs, such as visual images, could not be so validated because they lacked such unique associations with meanings or concepts. Yet those who have philosophically investigated the truth claims of modern language have warned against the eloquence of linguistic structures that may thwart the purposes of validation (Postman, 1985; Russell, 1966). Semiologists, such as Saussure and Pierce have further alerted us to the fact that words (signifiers) do not, and never did, have absolute or natural, unique links or associations to either the referrents (things that signifiers represent or refer to) or to the signified (concepts, mental constructs, and meanings elicited by the signifiers; Santambrogio & Violi, 1988). At the same time, with the growing recognition that, just like words, other signifiers (signs, symbols) are also subject to linguistic relationships (such as having syntax, semantics and pragmatics), it is increasingly realized that all symbolic systems have similar qualities and can be similarly used in transmitting information and knowledge. Contemporary semiotics has brought the realization that in all cases of signs or symbols, including the verbal, there are no stable or constant meanings or referentials, that all is in a state of flux. All meaning, therefore, whether words, sounds, visual images, touch, smells or tastes are used, is dependent on the interplay among signifiers which are only culturally attached to meanings.

Under the circumstances, therefore, substantive changes need to take place in education if we want the coming generations to be "literate" in the postmodern sense. (For an extended display of how terminology may have to change in postmodern culture, see Table 10.1). Just as schools have taught how to read and write in the past, there is a need now to teach *sensing* and *constructing,* to develop the abilities and skills in young people to "read" (sense) all kinds of sign(al)s they are exposed to from all sorts of

TABLE 10.1
Transformations of Literacy

Modern	Postmodern
Knowing subject	Communicating subject
Words	Signs
Objectivity	Immersion
Reading	Sensing
Writing	Constructing
Literate	Communicator
Literacy	(Multi)signefficacy

media in multilayered, multifaceted, multidimensional, hypertextual and hypermedia forms. Furthermore, a complete skill of *sensing* is dependent on a thorough understanding of how these sign(al)s interplay, intersect and interpret with and within culture(s). For such understanding and, thereby, complete literacy or (multi)signefficacy, all individuals must "write," that is, they must have extensive experience in *constructing* full-scale programs on latest media, such as the media that integrate the television, the computer and the telephone. Even children at early school age, for example, must be given the opportunity, indeed required, to not only use pen on paper—although this should still be an important part—but to take cameras, video equipment, computers, and so forth, and construct their own compositions (films, videos, music, virtual reality programs) that integrate all the sensual elements with all forms of sign(al)s. This may currently sound too optimistic, but unless it is exercised we shall have populations which have few literates who can *construct* (cultural, social, political) visions, and large numbers of illiterates who can only be an audience, that is sense in a passive manner.

CONCLUDING REMARKS

New information/communication technologies are raising new questions for human society, many of which may be directly related to the issue of literacy as discussed in this chapter. Consider, for example, the issue of democracy. Because literacy is linked to who, and to what extent, participates in the construction of the social imaginary and, thus, of the realities of social life an equal distribution of literacy seems to be a necessity for a democratic participation by all in the construction of social life. Those who are literate in the postmodern sense, explored in the previous section, will largely determine the conditions of life for humanity by constructing the hypes and the simulations; by communicating their visions of life effectively, and by signifying and representing the meanings attached to symbols through which a society makes life intelligible. Those who are not literate will be excluded from such construction. This seems to be especially true at a historical point in time when information has become the most influential source of power, when, many believe, those who control information control the social imaginary (Gitlin, 1987; Hanhardt, 1990; Kroker, 1992; Poster, 1990; Toffler, 1991).

Democracy does require rather constant and diligent vigilance on the part of all citizens. They have to be informed on many matters that influence their lives and constantly, because the speed and complexity of contemporary events create great dynamism. Unless there is genuine interest, willingness, and ability to be vigilant on the part of all, the dynamic changes will leave many on the outside. Many will not participate in the democratic process—meaning that the process will not be democratic. New communication technologies and media are widely criticized for distracting people from being informed and from intellectual discourse (Gitlin, 1987; Hanhardt, 1990; Kellner, 1990; Postman, 1985). They are blamed for promoting hedonistic, passive and "feel-good," enjoyment/entertainment-seeking lifestyles.

Blaming the technologies for such ills may be misdirected, however. As any other signifier, a technology is vulnerable to cultural signification. That is, technology, although maybe permitting certain uses with greater ease is, in the end determined by culture. Why a certain technology is selected over others, and the uses it will be put to are, that is, culturally determined. Take the example of television, a technology that has, without a doubt, had a major impact in contemporary society. When it was first introduced, and even still, it was (is) considered and promised to be a great educational tool. Despite its promise, it has been used and controlled first and foremost by commercial, advertising interests and purposes. Even in the case of news programs, as publicly stated by many television news producers (Moyers, 1989a, 1989c) and news anchors (for example, CBS news anchor Dan Rather), the commercial concerns override the informational content and value. Commercial sponsors and advertisers have, historically, constructed the meanings for and of television as a communication technology. As a result, around the world, television has come to be regarded as a medium of entertainment. That is how both the producers and the audiences of television programs largely approach their roles. Television is not a medium of intellectual discourse but a one-way relay station.

Clearly, this need not be the case, as exemplified by some highly educational and informative, discursive programs (on social events, nature, science, literature, etc.) found mostly on public television stations. When one encounters such programs, the futility of opposing books to television becomes most apparent. These programs can be as informative and intellectual as any book, with the added value of visual articulation and demonstration. Books can rarely match the intensity and documentation that a television program (a video) can provide in informing us of turtles' egg-laying habits or birds' migration patterns, for example.

Although modern society and the modernist rhetoric put greater value on books, clearly books can just as easily become "pure entertainment," as in the case of, say, Harlequin Romances. Of course, one may argue that there is intellectual information in every entertainment, and vice versa. In a "Seinfeld" situation comedy, for example, on television, or in a Harlequin Romance novel, there are attributions to and lessons to learn about feelings and approaches to life; and a good intellectual discussion, written or oral, in any medium, should to be highly entertaining. Consequently, it is all a matter of degree and utilization. It rarely makes sense, therefore, to apportion technology to knowledge or entertainment. Consider, for example, using virtual reality for educational purposes where students could be virtually transported to a historical period and experience "firsthand" historical events. Could not this immersed experience create greater interest and excite more reflection than a detached reading of history? As a matter of fact, especially if virtual sensual experiences are enhanced by verbal interjections, they may motivate students to explore more through all kinds of media, including books, and so forth.

The new technologies that integrate media are presenting us with great potential to integrate the best qualities of all. Will we take the opportunity? Will our culture allow significations that bring entertainment back into intellectual discourse and intellectual discourse back into entertainment? The answer to these questions may lie in who controls the significations: the corporations that consider the public as market and an audience, or the public itself. There are several attempts by people to allow public significations of these new technologies in order to make them means of conversation instead of one-way communication, as in the case of *Electronic Cafe International* (Galloway & Rabinowitz, 1989).

Will such efforts at democratizing new information technologies, and through them, our communities work? The response to this question is by no means clear. We are living a period when the commercial market with its well-ingrained economistic values and criteria of performance tends to fill in every void created by the disintegration of traditional institutions thanks much to popular disillusionment with the modern project and to postmodernist ideas. In the general speed of transformations and lack of postmodern literacy, the vigilance required for democratization may be lost. It is entirely possible, under the circumstances, that democracy, itself, may become a cultural commodity, not to be practiced as a way of life but to be (touristically) experienced as a consumable.

Many observers of contemporary cultures seem to agree that (post) modern society is increasingly one of speed (see, e.g., Poster, 1990; Toffler, 1991). Paul Virilio, a European philosopher with a theological background, argued that democracy has given way to *dromocracy* (see Kroker, 1992), that is, government by speed. Those who are the fastest will determine social choices and social reality (Toffler, 1991). Indeed, new information technologies introduce speed into all decisions at degrees unknown before, and this creates a further difficulty for human reason and for democracy. Speed allows little time or even possibility for reflection on affairs. Society is caught in an almost obsessive motion *forward*. Possibilities and potentials to do something new and exciting arise at unprecedented rates (speed) and there is little time to stop, take inventory, deliberate, converse, and reflect on the meanings and potential outcomes of these experiments. Also, social memory is lost since speed leaves little time for remembering the past. While the traditional political processes seem to have slowed down (even passing important legislation through Congress in the United States now seems so slow that by the time they are passed things already seem to have changed and people have lost patience and interest) everything else, technologies, cultural movements (Lacayo, 1994), lifestyles are changing with increasing speed. Consequently, there is a great need to embed into this speedy motion of events and changes the elements of *memory* and *reflection*. The new information technologies, especially hypertext and hypermedia, have a great potential for this embedding—but will it be done?

If not, Poster's idea of the computer, a machine, becoming the definer of the criteria and judgments through which society acts may actualize. Poster (1990) posited:

> Computer Science then is a discourse at the border of words and things, a dangerous discipline because it is founded on the confusion between the scientist and his or her object. The identity of the scientist and the computer are so close that a mirror effect may very easily come into play: the scientist projects intelligent subjectivity onto the computer and the computer then becomes the criterion by which to define intelligence, judge the scientist, outline the essence of humanity. (p. 148)

Of course, now it is not only the computer, but the computer integrated with the latest communication technologies. Thus, not only intelligence, but effective communication, a very important factor in construction of social reality may be technologically defined. Unless, therefore, the human is integrated into the technologies, he or she may well be left as the outsider.

REFERENCES

Angus, I. (1989). Circumscribing postmodern culture. In I. Angus & S. Jhally (Eds.), *Cultural politics in contemporary America* (pp. 96–107). New York: Routledge.

Baudrillard, J. (1983). *Simulations.* New York: Semiotext.

Body Digest. (1987). *Canadian Journal of Political and Social Theory, 11,* pp. 1–2.

Bolter, J. D. (1991). *Writing space: The computer, hypertext, and the history of writing.* Hillsdale NJ: Lawrence Erlbaum Associates.

Butler, J. (1990). *Gender trouble: Feminism and the subversion of identity.* New York: Routledge.

Butler, J., and Scott, J. (Eds.). (1992). *Feminists theorize the political.* New York: Routledge.

Derrida, J. (1982). *Margins of philosophy.* Chicago: The University of Chicago Press.

Eco, U. (1986). *Travels in hyperreality* (W. Weaver, Trans.). San Diego, CA: Harcourt Brace Jovanovich.

Ewen, S. (1988). *All consuming images: The politics of style in contemporary culture.* New York: Basic Books.

Feyerabend, P. K. (1981). *Realism, rationalism & scientific method: Philosophical papers* (Vol. 1). Cambridge: Cambridge University Press.
Feyerabend, P. (1987). *Farewell to reason.* London: Verso.
Fırat, A. F., & Venkatesh, A. (1993, August). Postmodernity: The age of marketing. *International Journal of Research in Marketing, 10*(3). 227-249.
Foster, H. (1983). Postmodernism: A preface. In H. Foster (Ed.), *The anti-aesthetic: Essays on postmodern culture* (pp. ix-xvi). Seattle, WA: Bay Press.
Galloway, K., & Rabinowitz, S. (1989). Welcome to "Electronic Cafe International." *Cyberarts,* 255-263.
Gitlin, T. (Ed.). (1987). *Watching television.* New York: Pantheon Books.
Greimas, A. J. (1987). *On meaning: Selected writings in semiotic theory.* Minneapolis: University of Minnesota Press.
Habermas, J. (1983). Modernity—An incomplete project. In H. Foster (Ed.), *The anti-aesthetic: Essays on postmodern culture* (pp. 3-15). Seattle, WA: Bay Press.
Hanhardt, J. (Ed.). (1990). *Video culture: A critical investigation.* Layton, UT: Peregrine Smith Books.
Harding, S. (1986). *The science question in feminism.* Ithaca, NY: Cornell University Press.
Jacobus, M., Keller, E. F., & Shuttleworth, S. (Eds.). (1990). *Body/politics: Women and the discourses of science.* New York: Routledge.
Kellner, D. (1990). *Television and the crisis of democracy.* Boulder, CO: Westview Press.
Kilbourne, J. (1987). *Killing us softly: Advertising's images of women.* Cambridge, MA: Cambridge Documentary Films.
Kroker, A. (1992). *The possessed individual: Technology and the French postmodern.* New York: St. Martin's Press.
Lacayo, R. (1994, August 8). If everybody's hip, is anyone hip? *Time,* pp. 46-53.
Lacoue-Labarthe, P. (1986). On the sublime. In L. Appignanesi (Ed.), *Postmodernism* (pp. 7-9). London: Institute of Contemporary Arts.
Landow, G. P. (1992). *Hypertext: The convergence of contemporary critical theory and technology.* Baltimore MD: The Johns Hopkins University Press.
Levin, D. M. (Ed.). (1993). *Modernity and the hegemony of vision.* Berkeley: University of California Press.
Lyotard, J-F. (1984). *The postmodern condition: A report on knowledge.* Minneapolis: University of Minnesota Press.
Lyotard, J-F. (1986). Complexity and the Sublime. In L. Appignanesi (Ed.), *Postmodernism.* London: Institute of Contemporary Arts.
Lyotard, J-F. (1992). *The postmodern explained.* Minneapolis: The University of Minnesota Press.
Marsella, A. J., Devos, G., & Hsu, F. L. K. (Eds.). (1985). *Culture and self: Asian and western perspectives.* New York: Tavistock.
Miller, M. C. (1989). *Boxed In: The culture of TV.* Evanston, IL: Northwestern University Press.
Most college graduates fail simple tests, study finds. (1993, December 17). *Los Angeles Times,* p. A41.
Moyers, B. (1989a, November 8). *The public mind: Image and reality in America—Consuming images.* New York and Washington, DC: Public Broadcasting Service.
Moyers, B. (1989b, November 15). *The public mind: Image and reality in America—Leading questions.* New York and Washington, DC: Public Broadcasting Service.
Moyers, B. (1989c, November 22). *The public mind: Image and reality in America—Illusions of news. New York and Washington, DC: Public Broadcasting Service.*
Nicholson, L. J. (1990). *Feminism/postmodernism.* New York: Routledge.
Poster, M. (1990). *The mode of information: Poststructuralism and social context.* Chicago: The University of Chicago Press.

Postman, N. (1985). *Amusing ourselves to death: Public discourse in the age of show business.* New York: Penguin Books.
Rorty, R. (1979). *Philosophy and the mirror of nature.* Princeton, NJ: Princeton University Press.
Russell, B. (1945). *A history of western philosophy.* New York: Simon & Schuster.
Russell, B. (1966). *Philosophical essays.* New York: Touchstone.
Santambrogio, M., & Violi, P. (1988). Introduction. In U. Eco, M. Santambrogio, & P. Violi (Eds.), *Meaning and mental representations* (pp. 3–22). Bloomington: Indiana University Press.
Sorkin, M. (Ed.). (1992). *Variation on a theme park: The new American city and the end of public space.* New York: The Noonday Press.
Steuerman, E. (1992). Habermas vs Lyotard: Modernity vs Postmodernity? In A. Benjamin (Ed.), *Judging Lyotard* (pp. 99–118). London: Routledge.
Toffler, A. (1991). *Powershift.* New York: Bantam Books.
Ulmer, G. (1990). *Art Papers, 14*(5), 17–22.
Williams, T. M. (Ed.). (1986). *The impact of television: A natural experiment in three communities.* Orlando, FL: Academic Press.

PART III
Policy Perspectives

CHAPTER 11
Telecommunications Policy and Economic Development: A Regulatory Perspective

James J. Malachowski
Rhode Island Public Utilities Commission

The relationship between telecommunications policy and economic development has been the subject of research for more than 25 years (Dordick & Wang, 1993; Porat & Rubin, 1977). This topic, however, has received renewed and quite intense scrutiny in recent years. This renewed attention has taken place in a variety of forums and has manifested itself in quite diverse arenas. In some quarters, the discussion has taken on an almost feverish pitch. It has been analyzed, of course, in academia, being the subject of numerous research projects (see, e.g., Dholakia & Dholakia, 1994; Dholakia & Harlam, 1994). It also has been debated in a more public and less scholarly way in a growing number of states by telephone companies, state legislatures, administrative branches, or utility commissions. The issue rose to newer heights when it was employed in a very effective manner during the 1992 Presidential election. The expression "information superhighway" became part of common language. Although most of the population is familiar with this phrase, few know what it actually means (Bryant & Love, this volume; Dennis, 1994).

In contemplating this issue, one immediately confronts a perplexing question. That is, does telecommunications modernization foster economic development or does increased economic activity cause telephone infrastructure investment? As the sophistication of research grows, many now believe the causal relationship, in fact, goes both ways (Dholakia & Harlam, 1994). Further research is exploring whether the causal relationship shown on the national level applies to the same degree on the state or county level. Various other aspects of this issue are also being explored (Andrews, 1994).

One difficulty in reasonably analyzing the causal relationship is that the arguments are being made by investor-owned, for-profit telephone companies that are employing this issue in hopes of gaining approval for major changes in the way they are regulated. Such changes often include a component that allows for the potential for greater profits.

For the purpose of this chapter, I write from the premise that telecommunications investment is important to a state's economic health. There may be lag of 2 years or more between network modernization and increased economic activity, and the relationship may occasionally be overstated due to the agenda of the presenter, but nonetheless, the conclusion here is that there is a relationship. Further, this relationship is significant and an important one (Richards, 1994). The goal of this chapter is to identify the policy issues that exist and offer a state regulator's perspective on the issues.

POLICY OR CONFLICT

There is an inherent conflict that exists in government concerning setting policy in the area of telecommunications. The problem stems from the fact that this is an area where the service is not provided for by government. For example, national and state transportation policies, which have so effectively developed the interstate highway system, were carried forward and implemented by the federal and state departments of transportation. Similarly, education policy is carried out by state and local units of government. The point is that government both creates the policy and government provides the service. Telecommunications, on the other hand, is provided for by public (investor-owned) and private for-profit companies.

Some postulate that the free market approach, deregulation, and increased competition should prevail in the telecommunications sector. After all, the free market approach is the backbone of our economy and our country. Several other countries around the world are abandoning other ideologies and moving to the free market. Since the late 1970s, the Executive branch has employed this type of laissèz-faire philosophy. Other countries such as Great Britain have also moved in this direction by privatizing state-owned or state-controlled industries, particularly in the utility sector.

During this same time period, however, we have seen developments and some advances in other countries that are employing a far different approach. In Japan, the government has committed to investing $120 billion to develop a digital broadband communications infrastructure. France has adopted a national policy that fostered the development of the Minitel system (Mansell, 1993).

Some in this country, after hearing of these developments, are calling for a stronger hand by the U.S. government in telecommunications. This implies a more centrally planned, government-controlled scenario. Under Clinton–Gore the federal policymaking environment has changed dramatically. This administration seems to believe the federal government should have a stronger role in the economy. This more centralized interventionist philosophy is manifested in the pronouncements about telecommunications. They believe this infrastructure is too important to the competitive position of this country to be left to the vagrancies of the free market. Clinton–Gore appear to favor an industrial policy that more closely links government with private businesses in key areas. Their concerns go well beyond setting standards and interconnection issues.

While government engages in the debate about its role, law and regulation impact the telecommunications sector more than any other U.S. business (Hontz, 1994). The entities involved include the FCC, the Department of Justice, the courts, the Department of Commerce, National Telecommunications & Information Administration (NTIA), state legislatures, state utility commissions, and local government. We currently have multijurisdictional, multilayered control, restriction, and regulation over this industry. In some cases, the authority of government agencies and the courts overlap. Figure 11.1 depicts the complex nexus of regulatory influences in this sector.

The valid and important question that is being raised is how much regulation and control is appropriate? How much of the free market approach should be allowed? How much competition exists or is likely to occur?

TWO MAJOR FACTORS: TECHNOLOGY AND COMPETITION

In this sector, what we are dealing with, whether it be voice, data, or video, is two people communicating; the rest is technology. Advances in telecommunications technology seem to occur daily and will go beyond the current digital switches and fiber optics. The technology is extraordinary and expensive. Determining useful lives, modernization schedules, and optimum platforms is no longer simply a depreciation-driven economic consideration. These factors are impacted directly and significantly by the second major factor: competition.

Corporate executives and public policymakers are driven by consideration of technology and competition that are the two major factors affecting the industry. The complexity is heightened in the situation where many believe that competition will only develop in some sections of the business (Kraut & Steinfield, 1994). Only certain customer classes will be served by a truly competitive market. Under this scenario, the dominant carrier has certain clear and strong incentives.

A major initiative of the dominant telephone company in this environment is to rebalance rates. The basic economic theory that state regulators are supposed to employ is that the cost-causer pays. The network and

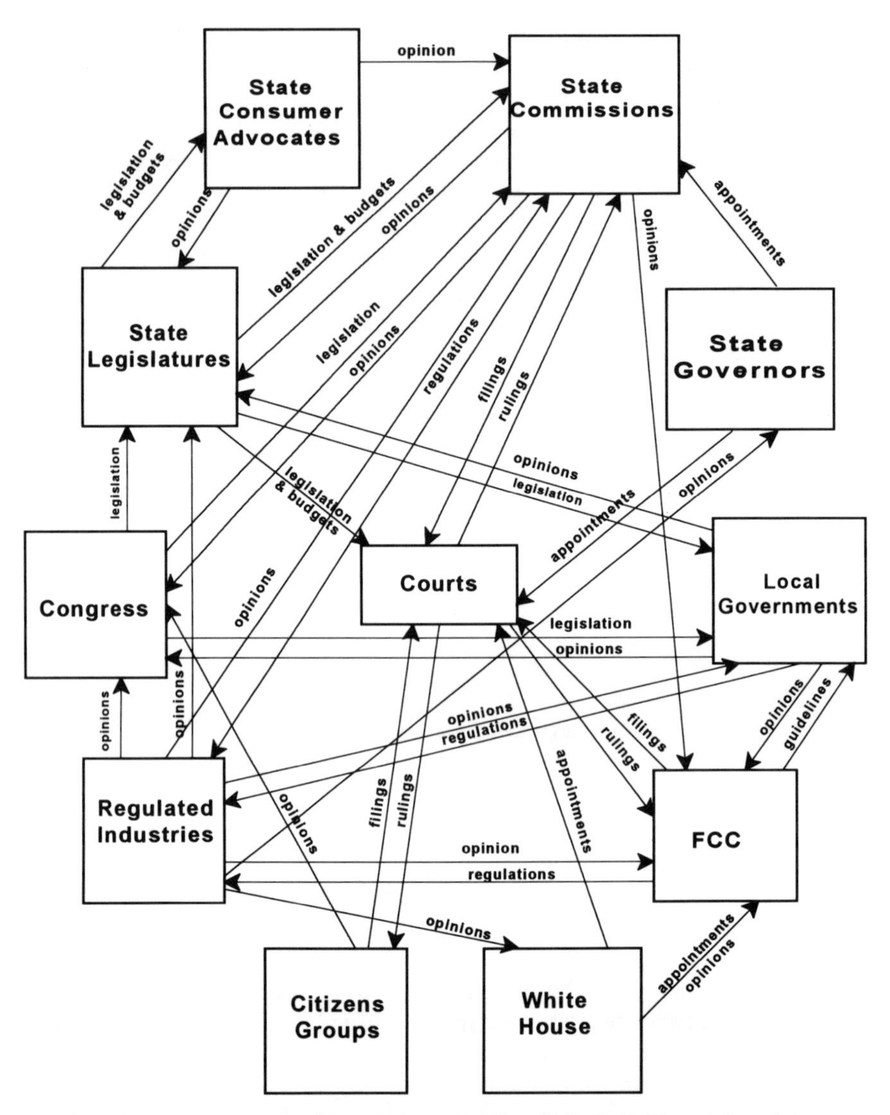

FIG. 11.1. Who decides what in communications policy? (Source: Adapted from M. Nodel "U.S. Communication Policymaking: Who and Where." *Hastings Communications and Entertainment Law Journal,* Vol. 13, Winter 1991, p. 290.) Copyright © Coral Weinhaus and the Telecommunications Industries Analysis Project Work Group).

administrative costs incurred by each class of customers should be assigned to and be reflected in the price paid by that class. Unfortunately, too often in the telephone area the prices do not reflect cost. The most descriptive example of this may be found in the additional charge for touch-tone service. Previously, in Rhode Island the price was $1.35 per line per month for residential customers and $2.02 per line per month for business customers.[1] The cost to the telephone company to deliver this product to a business or residence is not any different. The price structure was developed for one reason—to subsidize residential customers. This is a prime example of state utility regulators moving beyond their role of economic regulators and into the area of effectuating public policy.

When telephones were first being installed, there was a widely accepted policy of subsidizing the rates of residential customers. This policy was acceptable to the phone companies and their business subscribers because they shared the goal of ubiquitous telephone service. Customers could not call businesses if they could not afford to have phones in their homes.

In today's environment, local phone companies have strong reasons to rebalance rates and eliminate these cross subsidies. If competition will first surface in the high volume business area the incentive is to, in fact, shift cost to the captive residential customer. In considering rate rebalancing, regulators must ensure that the adjustments do not go too far and thereby injure vulnerable segments of society (Dutton, this volume).

A second goal of telephone companies is to obtain pricing flexibility, particularly for new products. In a competitive environment, a business must be able to respond quickly. The historic state regulatory process requires a tariff filing, a cost study analysis, a quasijudicial hearing, an open meeting decision, and the production of a written order. Pricing is a major component of corporate strategy. The ability to respond to markets and competitors is critical to implementing pricing strategy.

As technology evolves, phone companies will be offering a variety of new services. So called "phone-smart" services will bring an array of new offerings and other developments combining data, voice, and video. These new offerings will not (at least initially) be considered as part of basic service. Companies will be seeking freedom in pricing of these items.

A third goal of local companies that is driven by the dual imperatives of technology and competition is to be free from the constraints of traditional rate-based rate-of-return regulation and move to a more incentive-based scenario where prices are capped and earnings are unrestricted. Network modernization will distinguish a company from its competitors. New equipment, however, is extremely expensive. It is estimated to cost $420 billion to equip all customers in the country with broadband services. The concern in executive office is whether regulated returns based on the cost of capital will be a sufficient enough incentive to make the huge investments necessary to modernize networks in an increasingly competitive and, therefore, risky environment.

---

[1]The Rhode Island Public Utilities Commission (RIPUC), in Docket 1997, dated October 6, 1992, required that touch-tone become part of basic service and the additional monthly charge was eliminated for all customers.

COMPETITOR'S VIEWPOINT

An interesting and powerful component of state regulatory decision making that does not exist to anywhere near the same degree on the national level is the degree of accountability that state regulators are held to. In considering changes, some of which are major, a state regulator must deal with a skeptical public, a politically sensitive legislature, the image-conscious governor's office, concerned community leaders, alert rate payers, watchful consumer advocates, the state's attorney general, and others. The concerns are elevated when one is dealing with the telephone. This is due to the unique idiosyncracies of the consumer's relationship with the telephone. It is truly unlike any other utility sector that is regulated. There is quite a different feeling and attachment that people have for the telephone. Who really feels anything about or has strong opinions regarding electricity? Or water? Or gas? The uniqueness is further exemplified by the fact that despite good to excellent performance ratings by consumers for service quality, there remains a major distrust and perceived inefficiency attributed to the local phone company. Actions taken in this sector are scrutinized to a greater extent.

Arguments and positions of competitors also receive a wider audience. The telecommunications industry now contains a growing number of new players and new companies. These competitors argue that captive rate-payer dollars should not be used to fund network modernization. New services and advanced applications will only be of interest to a small number of customers in niche markets. It is not appropriate to force all rate payers to pay for technologies that will only be used by a few. It is not fair to mandate ubiquitous infrastructure deployments such as broadband or digitalization. Their position is that it would be more appropriate to have a slower market-targeted deployment plan. Further, phone company competitors argue that risk-taking entrepreneurs who will raise private capital are the more appropriate entity to deliver these services. They further argue that cost studies showing business subsidizing residents are flawed and pricing, particularly for new products, should be cost-based and subject to the rate setting process.

STATE REGULATORS' PERSPECTIVE

As with most problems, particularly complex and multifaceted ones, the first step is to break down the issues and develop some basic guiding principles. The RIPUC attempted to do this during its recently completed comprehensive review of the provision of telecommunications service. In doing so, five basic principles or beliefs emerged. They are:

1. Value of potential new telecommunications services to the community as a whole.
2. Technology hawk: The push for new technology.
3. Need to attract capital investment.
4. Ubiquitous deployment.
5. Regulation is flawed.

The remainder of this section is a discussion of these five issues.

Value to Community. In reviewing the first issue, which addresses the importance of the degree of sophistication of the telecommunications network to the community as a whole, one immediately confronts the economic development aspects (Dordick & Wang, 1993; Richards, 1994). As has been previously stated in this chapter, there is a converging belief that this is an important consideration. There is agreement that a causal relationship exists. However, having stated that, it is also believed this issue has been grossly overstated.

The economic development issue has been used as part of the strategy of local telephone companies to foster changes in the regulatory regime. This issue has been used very effectively when the proponents combine the intrigue of new technology (without having to explain it specifically) with a large dose of the issue that cuts right to the heart of the populace—jobs for the people and their children. The issue of economic development is also proffered during a period when the economy has been flat or declining in several sections of the country. The argument is presented and, at the same time, expanded to a much broader and far-reaching level—a state's competitiveness and, in fact, to the very "quality of life" of its citizens.

The debate is framed in the area of public policy. Some may find it strange how a for-profit firm, one that is publicly traded and, therefore, carrying stockholders concerns and goals, is able to legitimately argue that their interests lie in the area of public policy; how they can successfully argue their concern is the economic development and competitive posture of a particular state; how they can sincerely infer their capital investment decisions are affected by judgments and attempts to improve the quality of life in a particular area.

But this is exactly what is happening. Telephone companies have captured the high ground. They have cleverly moved the debate out of the hearing room. They have widened the interest and presented themselves in common, understandable language with broad appeal. They have upped the stakes by suggesting that it is not just the telephone companies' future that is at stake, but the economic vitality of the states and regions in which they operate.

State commissions that believe there is a causal relationship must be careful not to accept the extreme position of the phone companies. They must also address the key issue presented by the competitors. Should everyone be forced to pay for these new technologies or should deployment be user driven?

To help decide this point, one can again reference the earlier analogies of transportation and education. It has been determined that a comprehensive, safe, and modern highway system is important to the states and nation. Everyone pays for its development and maintenance. People have not been allowed to say, "I don't mind driving on bumpy roads, so I won't pay a portion of my taxes."

Likewise, it is an accepted policy that an educated populace is essential to the growth and well-being of the country. Everyone pays for education (often through local property taxes) even if one does not have children in school. These payments are substantial. In some municipalities, the education portion of the property tax bill is more than 50% of the total.

Does advanced telecommunications fall into the same category? This is the important public policy consideration. State commissions are being asked to move out of the narrow role of being economic regulators and into the realm of public policy decision makers to a degree not suggested in the past. In this light, one must review the basic premise offered by the local telephone company. Their arguments are built on the premise that change is necessary because there are growing competitive pressures. This, of course, means there are other businesses providing the same service. Therefore, it is appropriate to have only one of the for-profit businesses, even though it is the dominant carrier, be the entity that attempts to dictate public policy?

It is imperative that state commissions endeavor to open the public policy debate and make the discussion as participatory as possible. There are many voices that should and must be heard. Consumers, legislators, government officials, civic leaders, educators, low-income advocates, and the minority community are examples of the types of groups that should have a say in this debate. Special efforts must be made to reach out and receive opinion from these diverse groups.

After engaging in this type of process, the RIPUC decided that advanced telecommunications were important to the community as a whole.

Technology Hawk. In the midst of the great debate on telecommunications policy, an expression was heard that hopefully is not accurate but nonetheless is very telling. The statement was that in the telephone sector, if a piece of equipment was operating okay, it probably was obsolete! This phrase flies in the face of the high reliability of the network but was obviously meant to describe the rapid changes and advances that are occurring in telecommunications technology. As the sophistication of the network grows, it is hard to predict what the future usage and uses of the network will be. A prime example of this is touch-tone service. When it was first made available, touch-tone was promoted as a new technology that would make dialing easier. It was sold as an enhanced service or luxury item. Did anyone envision at the time of introduction that touch-tone would become a necessity due to such widespread use of the abilities it provides? It seems there are few places one can call without having to go through an automated answering system. There are also information access services such as financial information, stock prices, and sports scores all triggered via the touch-tone pad. These are just some of the growing number of uses for this technology that, in many jurisdictions, is being made part of basic service.

What will be the future uses of the new advanced infrastructure? Will some clever individual develop a mega application such as "windows" software was for the computer industry? Will it be developed by a 17-year-old in his or her garage?

Many believe this will happen. Numerous jurisdictions are endeavoring to make arrangements with telephone companies that call for the deployment of new technology. The RIPUC was very aggressive in pursuing new technology deployments but did so without dictating what platform or technology would be chosen.

Capital Investment. Although the Regional Bell Operating Companies (RBOCs) have successfully employed the economic development issue, this strategy has had a down side. The populist argument and positions of convenience used by telephone companies to advance their agenda have eschewed the debate and gotten off the stark economic reality of the situation. Twenty-four-hour, year-round service connected to every home and business with the "on-demand" requirement is expensive to build and maintain. The change out of the technology of these types of systems is, again, a very expensive undertaking. Large amounts of capital are needed to accomplish this task. Capital investments, in an increasingly competitive industry, raise the concern relative to the strength and degree of certainty of the revenue streams necessary to support the capital investment.

There is the core question that is faced, the real consideration we are confronted with. Will state regulators create an environment that encourages the necessary capital improvements and investments required to upgrade the public switched telecommunications network? Telephone companies and investors have a wide array of possible investment opportunities around the globe. If they choose to stay in their "line of business," many other countries do not have the restrictions the United States imposes. Focusing for a moment on capital invested in the domestic telephone system, Baby Bells operate in multistate territories. They have a wide mix to their franchise area that runs from highly sophisticated urban areas containing some of the major cities in the world to extremely rural, sparsely populated areas.

Capital allocation decisions are made on the basis of the greatest likelihood of achieving the targeted return on investment. When top executives of NYNEX, for example, meet to decide the capital budget for the states in their service territory, the conversation or decision does not focus on what state in the northeast needs the network modernization to assist its economic development. The allocations do not hinge on considerations such as: "We must invest in Rhode Island to make that state more competitive and improve its quality of life." The discussion and decision is centered on where is the greatest likelihood of gaining the required returns. Where can the most profit be realized? Competition brings uncertainty and greater risk. This is not the environment monopolistic utilities have historically operated in. The need to attract capital investment and establish a conducive environment should be a key consideration of state regulators.

Ubiquitous Deployment. When telephone usage was first being offered, the goal was to achieve as high a penetration rate as possible. Government and regulators accepted this as a matter of public policy. Pricing was structured in a way that the business sector subsidized the residential sector. The artificially low monthly rates for the residential customer helped speed acceptance. Later, a mechanism for subsidizing high-cost areas was developed. Subscriber line charges assessed on the long distance portion of the bill are collected from all customers and redistributed in the form of subsidies for local rates in high-cost areas.

These strategies have been helpful in attempting to achieve the goal of ubiquitous service. The penetration rate nationally is in excess of 90% (however, it is substantially lower than this for low-income and minority sectors). The focus of this issue has shifted. The concern is that if we are truly in the information age and telecommunications is the conduit for access, will all of the population have access? Will we all be able to participate in the information age? The most basic example is touch-tone service. If it is necessary to access information services and touch-tone is priced as an enhanced service requiring an extra monthly charge, those who cannot afford this cost will not have access. The more significant situation involves deployment of advanced network components such as digital switches and broadband fiber optic cables. These and other expensive network components will be necessary for delivery of advanced services. There is a growing concern that these features will be deployed selectively. Locations where businesses are concentrated and high income residential areas may not only be the first but may turn out to be the only areas where the advanced network is deployed. Many are concerned there will be an additional stratification in the country—between information-age "haves" and "have nots." This would not only apply to the residential sector. The backbone of the economy for several states (if not all) is the small business sector (Richards, 1994). Such businesses do not just exist in financial or business centers. These entities are spread geographically throughout the state. These smaller but important businesses should be able to participate in, and avail the benefits of, the information age. State commissioners, in considering the public policy question in the telecommunications area, must address this question.

Regulation is Flawed. Whenever government has its hand in a business sector effecting the operation of the free market, you have a flawed system. Regulation, however effective, will not produce the efficiencies that people want. RBOC responses to their changing environment has been interesting to a regulator. These companies are looking at their cost per access line and number of employees per access line in a way unlike ever before. Major workforce reductions are taking place at RBOCs in an effort to bring these ratios down. The press announcements issued constantly state that downsizing, process re-engineering, layoffs, or reductions are necessary to position the company for the current or pending competition. Years of regulation could not produce this result. The threat of competition has.

Rate-of-return regulation has shortcomings and deficiencies. Incentive regulatory schemes and price-cap models are also flawed. There is, however, one thing that is playing to the advantage of regulators. That is, they know the local phone company wants a change in the regulatory regime and they know the companies are willing to pay for the change. Companies are willing to negotiate and offer specific network modernization proposals and schedules along with price reductions or stability.

Depending on the aptitude of the state commission, deals are being made. New forms of regulation are being traded for new technology. At the beginning of 1994, 38 state commissions had approved some alternative form of regulation. One of the difficulties a regulator faces in contemplating this situation is that these new technologies bring with them a number of new telephone products. Caller identification, busy-line monitoring, and automatic call-back, call-trace, and ISDN are just a few. The telephone company is hoping these products will produce new and greater revenue streams. This is a different approach than the traditional one of a company attempting to increase market share. This strategy is to create new markets. Will these new products and new markets create new revenue streams that will substantially increase profits? While considering these issues and contemplating a change in the form of regulation a state commissioner, at times, may feel there is a wave of new products and revenues building up behind him or her getting ready to crash down, pushing company profits through the roof a year or two after the new regulatory regime is put in place. This would make it look like the regulator was fooled and tricked by the company and probably cost the regulator his or her reputation and job.

On the other hand, will these new sales simply offset the revenues the company loses to competition? This issue is certainly a perplexing one. Another difficult question is how does a regulator know that the phone company would not have made the network modernization investments regardless of regulatory change due to the company's need to stay ahead in the new competitive environment? A partial response to this is: How does the phone company know that the form of regulation would not have been changed anyway, regardless of the price and network modernization commitments in response to the new competitive environment?

CONCLUSION

It has been stated that the strategic resource of the first century of this country was land, for the second century the key resource was capital, and for the third century the strategic resource to create wealth and jobs will be information.

Much has been written and discussed about the "information age"—the effect it will have on our society; on the citizenry and upon our businesses. In the mid-1990s, more than 50% of all telephone traffic is data transmissions. Voice communications no longer accounts for the dominant usage of our telephone lines. The requirement is to move large volumes of data at high speeds. Traditional copper wire and analog switches do not have the capacity or bandwidth to accomplish this task.

It is imperative, therefore, for states to ensure they have the most advanced telecommunications infrastructure deployed in their jurisdiction. The best available telephone network is key for virtually every aspect of this country's economy. Companies that rely on information will be at a competitive disadvantage if they cannot count on their local network to deliver in a reliable and split second fashion. The biggest losers, however, will be individuals and small to medium-sized companies that cannot afford to supplement the public network with their own purchases of custom tailored equipment and software.

In 1984, Judge Green presided over the divestiture of the AT&T monopoly over the provision of telephone service. The break-up of local service into seven local operating companies, and the fostering of competition in the long-distance market, have brought significant changes to the telecommunications industry as we once knew it.

In the past few years, a combination of regulatory decisions, court rulings, technological innovations and changing markets have produced changes in the telecommunications industry which are as significant as those brought about by divestiture!

Examples include the Federal Appeals Court ruling that removed existing restrictions on telephone companies' provision of information services. This will result in deployment of electronic news services and yellow pages, health-monitoring services, and distance learning enhancements all being provided to consumers over telephone lines. In the regulatory arena, the decision to unbundle the network into discrete basic elements will allow independent enhanced service providers to build new services and to compete on an equal footing with the RBOCs. This is the so-called Open Network Architecture format.

The development and push for mass production of the so-called "smart-phone" is a technological breakthrough that will change not only what telephones look like but how we use them. This new phone consists of a sophisticated, intelligent computer video terminal with a touch-screen display that can simply "plug in" to a digital network. This new phone will work in tandem with an intelligent, high-speed, high-bandwidth, digitally switched, fiber optic transported telephone network; a network capable of simultaneously transmitting voice, data, video, and graphics over a single telephone line connected to the home or business.

In the face of a rapidly changing telecommunications industry, the regulators role has become increasingly difficult. No one can accurately predict the future but the imperative is for states to have an advanced telecommunications platform in place. This is necessary to be poised to take full advantage of whatever new communications techniques develop. Creative people will develop new and useful applications for the intelligent network. States cannot afford to fall behind in this critical area. Economic development, public policy considerations, capital deployment, technology, and competition are now part of the province of state utility regulators to a greater extent and to a more significant degree than ever before.

REFERENCES

Andrews, P. (1994, February 8). More realistic view of "information superhighway" is taking shape. *The Seattle Times,* p. D2.

Dennis, E. (1994). *Separating fact from fiction on the information superhighway.* New York: Freedom Forum Media Studies Center.

Dholakia, R. R., & Dholakia, N. (1994). Deregulating markets and fast-changing technology: Public policy towards telecommunications in a turbulent setting. *Telecommunications Policy, 18,* 21–31.

Dholakia, R. R., & Harlam, B. (1994). Telecommunications and economic development: Econometric analysis of the U.S. experience. *Telecommunications Policy, 18,* 470–477.

Dordick, H. S., & Wang, G. (1993). *The information society.* Newbury Park, CA: Sage.
Hontz, J. (1994, October 10). Hollings kills infohighway legislation. *Electronic Media, 1,* 2.
Kraut, R., & Steinfield, C. (1994, November). *The effect of networks on buyer-seller relationships: Implications for the national information infrastructure.* Paper presented at the Speech Communication Association annual convention, New Orleans, LA.
Mansell, R. (1993). *The new telecommunications.* Newbury Park, CA: Sage.
Porat, M., & Rubin, M. (1977). *The information economy: Development and measurement.* Washington, DC: U.S. Government Printing Office.
Richards, B. (1994, November 21). Linking up: Many rural regions are growing again: A reason: Technology. *Wall Street Journal,* pp. A1,8.

CHAPTER 12
Electronic Service Delivery in the Public Sector: Lessons From Innovations in the United States

William H. Dutton
University of Southern California

Visions of a national electronic network often lack compelling evidence about the public's need for or interest in this new age information highway. What services would be delivered over this superhighway of the information age? Will the public interest truly be served by its development?

Answers to these questions from the proponents of new telecommunication infrastructures are often quite vague. Frequently, the answers are reminiscent of overly optimistic forecasts of earlier decades in which many politicians, scholars, and journalists also foresaw a "wired nation."[1] Few decision makers are any longer willing to invest on the basis of optimistic scenarios about the public's enthusiasm for new electronic services. An increasing number of telecommunication experts simply appeal to faith in human ingenuity (Lucky, 1989). They argue that the future of information services is unknown, but that once new technologies are in place, the services will soon be invented—just as in *A Field of Dreams*—"They will come."

Many veterans of new media ventures since the mid-1980s, no longer

[1]Early developments of interactive cable television systems generated visions of a wired nation in which all kinds of information and communication services would be available to households and businesses over broadband cable networks (Dutton, Blumler, & Kraemer, 1987; Pool & Alexander, 1973). Similarly, the development of interactive computer systems generated visions of "public information utilities" that would provide citizens with public access via telecommunications to computers offering all kinds of computer-based information and services (Sackman & Boehm, 1972; Sackman & Nie, 1970). Many of the services envisioned for the wired nation and the public information utility are now being proposed for the fiber optic information highways of the 1990s, but proponents have not addressed the failure of these earlier visions of the future of communication (Dutton, Blumler, & Kraemer, 1987).

place much faith in the inevitability of new media services. The market failure of interactive cable television services in the 1970s and videotext services in the 1980s have led many erstwhile proponents of electronic services to conclude that electronic information services are simply not of interest to the general public. These critics argue that the proponents of the information highway have been oversold by the hype surrounding new media. The information highway is a utopian vision, or worse, simply a reinvention of videotext, that already failed in the early 1980s and will fail again in the 1990s. To these critics, the information needs of the public are already well met by existing mass media and telecommunication services (Forester, 1989; Noll, 1985).

It seems, however, that a case for electronic services might be emerging. Since the late 1980s, several dozen federal, state, and local government agencies as well as a number of quasigovernmental, not-for-profit agencies have been experimenting with the provision of electronic information services to the public. Their experiences provide one of the only empirically anchored perspectives available on the kinds of public needs and interests that could be served by electronic information services.

Over the last several years, my colleagues and I have surveyed many of these projects and spoken with key individuals involved with their design and management.[2] Our assessment of these projects underscores the diverse range of applications, technologies, and tasks that will define the future of electronic service delivery. They also provide an empirical basis for challenging commonly held assumptions about the future of public information networks and services. Taken together, these innovations in electronic services might indeed foreshadow a long-term trend toward the more direct provision of information services to citizens via broadcast, telecommunication, and other electronic media that are anchored in identifiable needs of the public. At the same time, these projects suggest that some visions of the future of electronic services are based on an overly simplistic view of the public's interest in information services.

In this chapter, I briefly survey the variety of initiatives underway in public networks and services in the United States. Based on the lessons learned from these initiatives, I then challenge five common assumptions about the future of electronic services and the public interest. I conclude by discussing the necessary conditions for an appropriate national policy for electronic service delivery.

[2]This chapter is based on research over several years with a number of students and colleagues at the Annenberg School for Communication. Additional reports of this research are provided by Dutton (1992a); Dutton and Guthrie (1991); Dutton et al. (1991); and Dutton et al. (1993).

INNOVATIONS IN THE PUBLIC AND NOT-FOR-PROFIT SECTOR[3]

A small but increasingly visible number of U.S. state and local government agencies as well as not-for-profit organizations are applying communication and information technology in ways designed to facilitate access to public information and services. These innovations find expression in a variety of different types of applications across a range of functional areas. Although there is no master plan shaping these individual initiatives, there are several threads that tie these efforts together.

One thread is an effort to emulate and catch up with the private sector on the assumption that the modernization of communication and information systems will bring major technical benefits in the speed, accuracy, and efficiency of public services. Another is a more or less explicitly articulated vision of using electronic media to bring government closer to the public. But the long-term, indirect consequences of these developments might have a relevance to politics and governance, which goes beyond improving the efficiency of existing practices, to altering the way we vote, poll, lobby, communicate, and obtain services.

Generally, innovative applications of communication and information technology fall within four general types of applications, described in Table 12.1, which my colleagues and I labeled as broadcasting, transactions, access to public records, and interpersonal communications (Dutton, Guthrie, O'Connell, & Wyer, 1991).

Broadcasting

The most prominent uses are aimed at broadcasting information for the general public. As cable systems have diffused, public agencies have more frequently used these channels for distributing government and other public information, such as by scrolling textual information or cablecasting live events, such as council meetings. Cablecasting of governmental affairs is becoming a common practice, but localities like West Hartford, Connecticut have advanced the state of the art in this area by producing more sophisticated local news and public affairs programming and committing themselves to the concept of "neighborhood TV" (Dutton et al., 1991). Rather than simply pointing a camera at live events, these producers and growing ranks of volunteers seek to translate lengthy meetings into bite-size programs that capture the interest of the local audience. Old visions of

[3]The findings discussed in this section are adapted from a survey conducted by the author along with Kendall Guthrie, Jacqueline O'Connell, and Joanne Wyer (Dutton et al., 1991).

TABLE 12.1
Emerging Electronic Services Supporting Citizen Access[a]

Communication Task	*Systems Employed*	*Early Applications*	*Selected Examples*
Broadcasting	Satellite-linked and local cable TV systems	Cablecasting of public meetings, hearings; neighborhood TV	C-SPAN, California Channel (Cal-SPAN); West Hartford, Connecticut
	Electronic bulletin boards	Continually updated information on events, agendas, services	Santa Monica's PEN, Pasadena's PARIS/PALS
	Touch screen, multimedia PC	Multilingual kiosks	24-hour City Hall, Hawaii Access, LA Project, L.A. County Library Pilot
Transactions	Touch screen, multimedia PCs or kiosks	Apply for social services	Tulare Touch
		Renew/update driver's license	California DMV's Info/California
	Magnetic strips and smart cards	Electronic benefit transfers	New York City Food Stamp benefits
	Voice processing	Schedule inspections	Arlington County
	Electronic mail	Requests for services, completion of applications, licenses	Santa Monica's PEN
	Automated teller machines (kiosks)	Welfare and medicaid transactions	Proposals at the state and federal levels
	Automatic vehicle identification systems	Automated assignment of tolls for use of roads	Under discussion in California
Access to public records	Audiotext, recorded messages	Answers to routine public inquiries	Hillsborough County's Fact Finder; Phone Phoenix
	Dial-up electronic bulletin boards	Public, government, community information	PARIS/PALS Pasadena; Santa Monica's PEN; New York City Board of Education's NYCENET
	Dial-up access to public databases	Access to land records by title companies	Proposed in Arlington County, Virginia

Interpersonal communications	Expert systems to assist information providers	Assist information & referral providers in offering advice & referrals	Under development by INFO LINE in Los Angeles
	Voice mail, asynchronous voice communication	Parent–teacher exchanges	City of New York
		Inquiries about teacher certification	State of Connecticut
	Facsimile	Application for building permits	City of Spokane
	Electronic mail	Citizen complaints, inquiries, requests	Santa Monica's PEN
	Computer conferencing & bulletin boards	Electronic forums on public issues	Santa Monica's PEN; NYCENET, FreeNet
		Electronic networking to support groups & voluntary organizations	Santa Monica's PEN, HandsNet, Community Link, Reference Point
	Audio and video conferencing	Arraignment; bond reviews; and other pretrial meetings	San Bernardino, CA; Dade County, Florida, . . .

Note. Dutton et al. (1991) and Dutton (1992b).

community programming over cable are beginning to be more closely approximated in a few localities.

Beyond the local level, the success of the Congressional-Satellite Public Affairs Network (C-SPAN) has prompted state officials to emulate their programming (Westen & Givens, 1989). The State of California has launched a system, which it calls the California-Satellite Public Affairs Network: The California Channel (CAL-SPAN).

In the early 1980s, some laboratory and field trials of teletext were launched to explore the value of teletext as a means for broadcasting public information (Elton & Carey, 1984). In the early 1990s, state and local governments are experimenting with public information kiosks, but designed around multimedia personal computer platforms. These kiosks generally combine laser disk storage devices and microcomputers with a touch screen to allow people to access government and other public service information. They are being placed in libraries, shopping malls, recreation centers, and grocery stores to provide public information in the form of text, video, graphics, or sound, all in multiple languages to reach a more multicultural society. They are promoted as a means for bringing information closer to communities, not expecting citizens to travel to city hall or speak English.

In this area, Public Technology, Incorporated, in partnership with IBM has supported a "24-hour City Hall" project, which placed kiosks in more than a dozen local jurisdictions. In 1992, immediately after the Los Angeles riots, North Communications, a Los Angeles firm that has worked with IBM in developing kiosk systems, launched what it called the LA Project. North distributed a small number of kiosks throughout Los Angeles, designed to support South Central and neighboring Los Angeles residents in seeking phone numbers, public information, business loans, employment, and other assistance of possible value to recovering from the urban violence that followed the Rodney King verdicts.

Transactions

The Department of Social Services in Tulare County, California has developed a kiosk system called "Tulare Touch." Tulare Touch uses a multilingual, video and audio touch screen connected to an expert system. The department has 35 kiosks with a touch screen at each district office. Welfare clients step up to a kiosk, choose which language they wish to use, and are walked through a welfare application. Fraud screens provide checks of income or other information by linking with other files to verify the client's eligibility. The system approves or rejects the applicant at the time of application or gives the applicant a list of information needed for any future application. The project was aimed at reducing costs by using the capabilities of an expert system to minimize errors in determining eligibility and calculating payments.

In late October 1991, California's Department of Motor Vehicles (DMV) launched a 9-month pilot project called "Info California" (Dutton et al., 1991). Fifteen information kiosks were initially placed in Sacramento and San Diego to pilot this experiment, permitting citizens to conduct a variety of transactions, such as registering an out-of-state vehicle. The system was developed by IBM and North Communications. This system was modeled after an automated teller machine (ATM).

One major thrust of the Info California project is to overcome the jurisdictional fragmentation of services and permit the public to access information at any level of government through a single facility. The director of California's Health and Welfare Data Center wishes to "provide a single face to government for the citizens of California" (Hanson, 1992, p. 16). Various agencies in California, such as the Los Angeles County Library, which have had some success with standalone projects, are actively considering the use of the Info California system as a platform for the wider offering of their services.

Other state and local governments are using similar electronic systems for distributing benefits, such as food stamps, hoping to provide recipients with benefit cards that could be used much like a credit card in local groceries and supermarkets. Other government agencies are experimenting with smart cards as another approach to automated payment and billing.

Arlington County, Virginia has implemented what is called the "parkulater," a device about the size of a pocket calculator. It uses a small computer chip and acts as electronic dollars for commuters parked at meters. The user types in the type of meter, places it in the car's front windshield, and the time runs until the driver leaves and turns it off. Drivers are billed for time they are actually parked and they do not need to have the correct change. By 1992, as many as a half dozen other jurisdictions were using this technology and already beginning to face problems with a lack of standards for parking fees across neighboring jurisdictions.

Access to Public Information and Records

New technologies are being employed to provide an additional means for providing public access to information and records. The technologies employed range from audiotext, menu-driven telephone systems, facsimile, computer bulletin boards, and videotext, to multimedia personal computer kiosks. All are designed to supplement rather than replace more conventional systems for accessing public records over the phone or over the counter.

Voice mail systems are being used to answer routine citizen inquiries.

"Phone Phoenix" is one system offering prerecorded messages over the telephone as a means for answering many common, repeated questions and disseminating information, such as a schedule of current events. The system records menu selections made by the public, enabling the city to determine what topics are of most interest to citizens. Some cities are using these systems to handle more specialized information needs. Dallas, Texas uses audiotext to communicate with people requesting information about the status of building permits. Arlington County has implemented a form of audiotext that provides building contractors dial-in access for scheduling a building inspection.

Like many other telephone-based information and referral agencies, INFO LINE, a nonprofit California corporation, uses a menu system to route calls to the appropriate information specialists, such as those focusing on emergency food or emergency shelter. Interestingly, the major innovation at INFO LINE in recent years has been a move toward the establishment of neighborhood centers, which can provide a walk-in service and also locate small service providers within a neighborhood that would not otherwise be visible to their centralized information and referral database. In Los Angeles County alone, INFO LINE maintains information on more than 5,000 service providers. Its first two neighborhood centers were established in a Korean-American community and in a distressed African-American community of South Central Los Angeles.

A number of jurisdictions have experimented with electronic bulletin boards and other online systems for offering dial-up online access to databases. Libraries have been among the most likely agencies to develop online services, allowing computer users to access the library's card catalogue from their home or office. The public library in Pasadena, California expanded on this concept to develop a database of civic organizations, city commissions, and social service agencies as well as a community calendar listing everything from performing arts programs to garden club meetings; they call the system Public Access Library System (PALS; Guthrie, 1991). The library of the City of Glendale, California plans to launch one of the most technologically state-of-the-art public access computer systems, which they call the Local Information Exchange (LNX).

A few other cities are using computer bulletin boards to provide citizen access to information. The City of Santa Monica, California has led the way with its Public Electronic Network (PEN). PEN contains information on community events, the council's agenda, as well as other information on city departments, officials, and services. It is accessible from personal computers or any of 25 public terminals distributed throughout the city (Dutton & Guthrie, 1991). The New York City Board of Education has also developed a computer bulletin board called NYCENET, used by both educators and students.

A number of government agencies are considering the provision of online access to specialized databases, both as a means for improving services as well as a way to generate new revenues. For example, some agencies, such as in Arlington County, have considered providing of online access to land records; they believe some title companies might be willing to pay for the convenience of accessing these records from the company's offices. If land records are automated for other reasons, this kind of service is increasingly practical at only a marginal cost. Hampton, Virginia has provided 25 incoming lines to their computer, where people who subscribe can access information such as the sales history of property, the owner, city tax assessments, school assignments, and permits. Geographically based information systems that integrate census, housing, and other information are another specialized database that is potentially of value to telemarketing firms.

Online access to records, increasingly stored on optical disk imaging systems and integrated with facsimile equipment, provides an attractive alternative for private firms to use in accessing records and a promising new revenue source for the public sector. Information can be retrieved for on-screen viewing or directly "faxed" to the user.

Interpersonal Communications

Government officials and personnel have long communicated with the public in person, over the counter, by mail, and by telephone. Increasingly, they are using voice mail, facsimile, electronic mail, and computer conferencing systems.

Information and referral agencies rely very heavily on interpersonal communication between the public and information specialists. The information specialists often provide counseling and emotional support as well as information to their clients. One important innovation in information systems is designed to support the communication between client and information specialist by providing the information specialist with an expert system to assist in the identification of appropriate agencies for referral. As being implemented by INFO LINE in Los Angeles, the information specialist will input data on the client's location, age, gender, any handicaps, and needs. The system will identify the most proximate and relevant agencies for referral. Eventually, this system will replace a heavily annotated manual file of information on 5,000-plus agencies that sits in front of the information specialist. It should speed and improve the delivery of information to clients and overcome some of the inherent limitations of a service that is too heavily dependent on the memory of specialists.

Voice mail is viewed as a means to reduce the staff time required to handle routine telephone calls. It is also being used to enhance rather than replace interpersonal communication, for example, by supporting parent–teacher communication, allowing the teachers to leave a message at the end of each day and providing the working parents with a means to hear the teacher's message or leave a message for the teacher. Other state and local agencies are using voice mail for answering inquiries about teacher certification, filing insurance claims, and inquiring about the status of claims.

Cities are also using facsimile machines to receive and, more recently, distribute forms. Cities can send and receive applications for permits via facsimile. Optical disk storage systems can be used in conjunction with facsimile machines to distribute records identified through an online search.

A potentially more dramatic departure is the use of electronic mail. This technology has been adopted since the 1970s by government agencies for internal, organizational communications, but a few governments have begun to apply this technology as a means to support dialogue with the public.

The City of Santa Monica's PEN stands out in this area. PEN allows citizens to send electronic messages to city hall, council members, any city department, and the library's reference desk. City staff can then reply to the request electronically. City departments received more than 5,000 messages from the public in the first 2 years of PEN's operation. Sixty percent of the individuals described their messages to city hall as inquiries. About one fourth sent either comments, complaints, or other kinds of messages to city officials (Dutton, Wyer, & O'Connell, 1993).

Some city department heads and supervisors in Santa Monica were reluctant to implement electronic mail between citizens and city hall for fear it would open a floodgate for complaints. In PEN's case, it does seem to have increased the number of complaints and requests coming to the city. Therefore, this facility is perceived to have increased the workload on city staff but also the responsiveness of the city to the general public (Dutton et al., 1993).

PEN is also innovative in its provision of facilities for citizens on the system to send electronic messages to one another as well as to the city. This augments PEN's use for electronic meetings and conferencing about public affairs. In Santa Monica and elsewhere, electronic mail and conferencing software are being used to support public forums on policy issues. In contrast to electronic voting and polling systems proposed in the 1970s, these forums are designed to foster dialogue to set agenda and explore options rather than facilitate voting and polling from the home (Dutton et al., 1991). Although Santa Monica's PEN system is the only municipally funded project that has extensively developed a conferencing capability, political forums have been developed on Cleveland FreeNet, a nonprofit operation in Cleveland, Ohio, and by Berkeley's Community Memory project. The New York City Board of Educations' NYCENET System also features discussions (Dutton et al., 1991).

Finally, government agencies are using audio and video teleconferencing in creative ways to reduce travel and increase communication with remote locations. One instance is in the criminal justice area, where a number of jurisdictions are trying to reduce their costs for transporting inmates by providing a two-way video link between their jails and courts. People arrested for misdemeanors, and in some jurisdictions for felonies, can be arraigned and receive a bond review from the jail to which they are taken after their arrest.

There are a variety of other applications of electronic media that are increasingly moving citizens in more direct, albeit mediated, interaction with government agencies. Advances in computer-assisted dialing and interviewing systems are being used by public opinion polling firms and could be used by governments in similar ways to gather information about the public. The entertainment media are experimenting with new versions of interactive television that utilize over-the-air broadcasting, in home terminals, and the telephone network to poll viewers as they play along with televised game shows. The same technology could be used by government agencies to interact with the audiences of public or governmental affairs programming, much as they did in Columbus, Ohio when the QUBE system, developed by Warner-Amex in the early 1980s, was still in operation (Davidge, 1987). Such systems could and have resurrected interest in voting and polling from the home, even if it is problematic that they will ever diffuse to a larger proportion of the public.

Also, the ordinary push-button telephone could be used increasingly for polling citizens on a variety of issues—either alone or in conjunction with televised speeches or debates. The introduction of screen phones, initially being designed to facilitate voice processing by presenting options over a liquid crystal display device mounted on the phone, will undoubtedly encourage such applications.

Finally, I have not mentioned the use of video and other electronic systems for monitoring the public, even though they are another application that will increasingly bring the public in more direct linkages with governments. For example, transportation agencies plan to employ video more routinely as a means to monitor traffic conditions in real time. Caltrans, the California Department of Transportation, is developing an automatic vehicle identification system (AVI), much like that developed in Singapore, to automatically charge for tolls. Singapore uses its system to control traffic by charging additional fees for the use of streets within the central business district during peak rush hours.

EMPIRICAL CHALLENGES TO COMMON ASSUMPTIONS

In light of some of the lessons learned from the innovations discussed here, it is possible to take a critical look at some common assumptions of both the critics and proponents of public access to electronic services. From the perspective of developments underway, primarily in the public and not-for-profit sectors in the United States, it appears that the prevailing assumptions oversimplify an increasingly diverse set of developments that are reshaping our notions of public access to information. Both the critics and proponents of electronic services could benefit from a more systematic look at the evolving nature of these services.

Electronic Service Delivery Is Futuristic

Since the 1960s, when the visions of a wired nation and public information utilities were first promoted, the most common criticism of the information highway has been that it remains a utopian pipe dream. However, aspects of these early visions are no longer so futuristic or utopian as they once seemed (Williams, 1991). Unlike discussions of the public information utility in the 1960s, for example, public electronic service delivery is a real development with many concrete, albeit emergent, forms and varieties. As just described, a diverse array of electronic information services are emerging to support citizen access to information and services.

New electronic services entail the use of computers and other electronic media to broadcast information, conduct routine transactions, provide access to public records and information, and communicate with the public. The technologies that will support and extend these applications, such as multimedia personal computers, screen-based telephones, and facsimile machines, have already arrived. Tulare County is already using a touch-screen terminal for clients to complete intake forms. Ross Perot and others have already conducted electronic town halls of sorts. Santa Monica already has electronic forums on public issues via the PEN system. The White House is online.

It is no longer a question of whether the public information utility will become a reality, but how and in what specific forms. In this respect, public information utilities are emerging in a form that is technologically more diverse, organizationally more decentralized, and functionally more focused on communication than anticipated by the early promoters.

We Haven't Learned From the Failure of Videotext

The failure of commercial trials of videotext systems in the early 1980s convinced many telecommunication experts that the public is simply not interested in having electronic information at their fingertips. As A. Michael Noll (1992) at the Annenberg School for Communication at the University of Southern California put it: "Videotext by any other name will also be a failure."

It is true that hybrid forms of videotext systems are among the initiatives being undertaken in the early 1990s. Multimedia information kiosks as well as many electronic bulletin board systems are based in part on a logic very similar to videotext—which is that the public is interested in electronic access to information. However, these are only part of a more general and diverse experimentation with electronic information and service delivery technologies of all kinds. No single model describes current developments—not videotext, not cable TV, not the personal computer, not an automated teller machine, nor a telephone. Taken as a whole, these initiatives underscore the specialized uses and myriad hybrids of all of these systems rather than any single dominant technology of choice.

More specifically, there are several features of emerging services that distinguish these systems from videotext trials of the early 1980s. Technologically, anyone with a personal computer who wishes to use a bulletin board would only need to purchase a modem and communications software to dial into an electronic bulletin board. In fact, anyone with a personal computer and modem could even set up an electronic bulletin board with software that is available free as "shareware." These incremental costs are substantially lower than those involved in purchasing or leasing a specialized videotext terminal and paying for the use of these services. In many cases, access to information on a public kiosk or system like PEN is free to the public, unlike early videotext systems, which required users to pay a variety of charges. Moreover, the public systems are not directly subscriber-supported, and they are geared to a specialized rather than a mass audience.

The underlying technology of the newer services is far more specialized and specific to the particular application than were videotext services, which sought to use the same technology to meet every need. No public agency, for example, is employing a specialized videotext terminal, such as those used in the Times-Mirror Gateway or Knight-Ridder videotext trials, although a number offer electronic access to information via terminals or personal computers. Unlike early videotext trials, agencies are employing cable television, satellites, multimedia personal computers, facsimile machines, and ordinary telephones. Audio is being employed as often as visual information when audio seems sufficient, as in many information and referral services.

Organizationally, these systems are often launched by voluntary, not-for-profit, and public agencies seeking to address specific problems with access or delivery of services. They are problem-driven. The systems are not being launched by newspapers pursuing a defensive strategy to compete with would-be competitors.

Finally, emerging applications are designed to support a more diverse range of functions than earlier videotext systems that primarily focused on the provision of information. Emerging systems are often more geared to support communication and transactional services than to the provision of information per se.[4] Instead of simply putting information at your fingertips, the catch phrase of new developments might be putting government at your fingertips.

The Information Needs of the Public Are Well Met by Existing Media[5]

Many computing and telecommunication projects seem to be driven by technological possibilities rather than by public needs. In theory, the interests and needs of the public should guide technological change. In practice, it is difficult to identify needs far enough in advance and in clear enough terms to shape the design and development of technology. However, our survey of innovations in electronic service delivery along with more focused research on electronic service delivery in the inner city surfaced a variety of public needs and interests that are embedded within the design and development of innovative electronic services (Dutton, 1992a, 1992c).

My colleagues and I found quite genuine needs among the public for information services, whether or not they are driving the development of electronic services. Broadly, they can be classified into two general types of needs: access to information and access to information technology, the new medium for obtaining information and communication services.

Access to Information. In an era of so-called information overload, few managers or professionals can imagine situations in which there is truly a

---

[4]Early studies of videotext discovered that information retrieval incorporated a limited view of the functionality of this technology. Transactional and communication services were judged to be far more central than suggested by early visions and commercial trials of the videotext (Hooper, 1985; Noll, 1980). The experiences of state and local governments seem to be rediscovering the marginality of information as compared with transactions and communication functions (Dutton et al., 1993).

[5]This section is based on a report written for the Office of Technology Assessment (OTA), U.S. Congress, on a workshop contributing to OTA's study of federal telecommunications in the 21st century, see Dutton (1992c).

lack of essential information. Ironically, that is precisely the case, particularly among the less well-to-do.

For example, although it is well known that many people go without food or shelter due a lack of a sufficient number of food programs and shelters, it is less well known that some individuals simply do not get needed food and shelter for the lack of information on its availability. INFO LINE, a nonprofit agency, offers public information and referral services over the telephone to anyone calling within the County of Los Angeles. They receive from 220,000 to 230,000 calls per year and have been processing about the same number of calls over the last 3 years, despite deepening problems with unemployment, homeless, and public services within the County and City of Los Angeles. One of their managers argues, convincingly, that the number of calls they can process is primarily a function of staffing. If they doubled staff, they might be able to double the number of calls they process.

Currently, there is an average wait of 7 or more minutes per call before an information specialist can speak with the client. They have no way to track the number of people who hang up before their call is answered, but they realize it is a high number. Given that the two most frequent calls to the agency are for emergency food and emergency shelter, a failure to get information is potentially depriving individuals of a meal, a place to sleep, or other basic needs.

Other examples of information bottlenecks exist. One that potentially affects every citizen is the processing of 911 calls in major cities, including Los Angeles. According to some authorities, it takes up to 10 minutes to answer some 911 calls in the city of Los Angeles. One survey of over-the-telephone operators in both the private and public sector found that one of their most constant frustrations is apologizing to callers who are continually upset because they have not been able to get through on the phone (Fountain, Kaboolian, & Kelman, 1992).

A parallel can be found in the area of broadcasting where there are also needs in the midst of abundance. Many ethnolinguistic minorities in the United States are poorly served by the mainstream news and entertainment media. For example, in the disturbances in Los Angeles following the verdicts in the Rodney King beating, the Korean-American community was very dependent on local Korean broadcasting, primarily radio, and local Korean newspapers to keep their community informed about the course of events.[6] In the hours after the disturbances broke out, Korean-Americans quickly concluded that the mainstream media did not know what was going on. There was a clear sense within the Korean-American community that

[6]This discussion is based on a presentation by Kay Kyung-Sook Song at an Office of Technology Workshop on Electronic Service Delivery and the Inner City (Dutton, 1992c).

Korean-American shops were being targeted by looters, but there was no mention of this possibility on the mainstream media, which set the agenda for the police and politicians within Los Angeles.

Working with Korean broadcasting, leaders within the Korean-American community were able to marshall enough evidence to demonstrate their case to the mainstream media. Eventually, coverage by the mainstream media captured the attention of politicians and police agencies, who then directed more resources to help Korean-American and other small shop owners.

This case is but one example of an enduring problem of encouraging minority-oriented programming by commercial broadcasters. Although critics have long noted the need for minority-oriented programming for television, the equivalent need for minority-oriented software for computer systems is only now emerging on the public agenda. Software for a personal computer is quite analogous to videocassettes for a television equipped with a videocassette recorder. Similarly, computer software that is an increasingly significant commercial product is as likely to be oriented to the mainstream culture as is a television program.

Cultural barriers are tied in a variety of ways to educational barriers. Education has long been viewed as a barrier to access for minorities. Skills in using many technologies are supported by traditional education, specifically reading, writing, and arithmetic, which offer a gateway to functional literacy within a variety of areas. Most computer software and its associated documentation assume at least a high school education and knowledge of the English language as well as a background within the mainstream culture.

These cultural and educational barriers create a general need for bilingual staff and multilingual services tailored to the diversity of languages and cultures that compose our society. The Los Angeles County Library offers services to over 45 different language communities (Dutton, 1992a). The L.A. Project, which provided information on about 200 social services over multimedia personal computer kiosks, offered three language options—English, Spanish, and Korean—with some success.[7] Commercial services, not necessarily developed with distressed populations in mind, such as AT&T's translation services, have often been used by information specialists at INFO LINE, Los Angeles County's information and referral agency, when they encounter a language they are not prepared to serve.

Multilingual services require more than simply offering services in more than one language. Every service needs to be viewed from the perspective of a non-English-speaking client. For example, INFO LINE found that many Spanish-speaking clients would hang up if they heard a message recorded in

[7]Michael North noted that about 100 people used each kiosk per day for about 8 to 10 minutes per person (Dutton, 1992c).

English before they could hear the message repeated in Spanish. INFO LINE experimented with some success in switching the order of presentation, offering the Spanish instructions first. Similarly, their staff found that a flyer inviting individuals to call "213-OPEN" simply confused their Spanish language clients.

Political jurisdictions create another barrier to public access and, therefore, a need that might be addressed by electronic service delivery. In experimenting with a touch-screen information kiosk, staff of the Hall of Administration of the County of Los Angeles saw immediate disadvantages in the fragmentation of services among multiple departments and jurisdictions. The public did not want to move from kiosk to kiosk to find information or services.To create a successful information kiosk, it seemed important to provide an ability to cut through the boundaries of departments and jurisdictions so that the public could find the information they sought whether it was collected and maintained by the State Department of Motor Vehicles, the County Hall of Administration, or the City Attorney's Office. Just as the banking industry has permitted access from nearly any automated teller to nearly any bank or credit card company, despite the firm or location, so should a public kiosk provide access that overcomes the institutional fragmentation of the public sector.

One of the most formidable barriers to creating truly valuable information systems for the public is cost. Anyone who has even set up an electronic directory of addresses and phone numbers quickly learns that the development of a complete and up-to-date database is a major investment of time and energy that must be constantly attended. North Communications was able to set up six touch-screen information kiosks around South Central Los Angeles within 12 days after the disturbances following the King verdicts. The kiosks contained information on about 200 services. Among the lessons learned by North Communications was that "getting enough information into the system quickly enough" was one major requirement. They also discovered that the data was highly volatile and that they needed to make frequent updates.

North Communications was not alone. According to one of the top managers of a Los Angeles information and referral agency, in the aftermath of the disturbances in Los Angeles, "many agencies and elected officials felt that they had to establish information lines, but few of the lines they started had much value" (Wallrich, personal communication, 1992). As he put it: "A needless proliferation of information services is not good at any time, and especially should be discouraged during a crisis, when inaccurate information can be very harmful."

Finally, the harried pace of working families often makes it difficult for individuals to attend meetings during the day or evening hours, which lessens the adequacy of traditional approaches to informing the public. Our survey of innovative projects highlighted a number of efforts aimed at addressing a perceived decline in their ability to reach "live audiences." Neighborhood television projects, such as that in West Hartford, Connecticut, are not meant to substitute for attendance, but to provide some opportunity for residents to be informed when they would otherwise not be able to see the event at all. Likewise, a sizeable number of PEN users are active in local political discussions by virtue of their access to PEN, when otherwise they would not have been involved in public affairs because their family or work requirements would not permit them to attend regularly scheduled meetings.

Indirect evidence that electronic network services might diminish some barriers to communication between the public and institutions emerged from our research on PEN. We surveyed personnel in the City of Santa Monica about their perceptions of the impacts of PEN on their work and the responsiveness of the city to the general public. The general pattern of their responses provides a convincing case that the city's personnel believe that they are getting more requests for service, face more time pressure, and have become more responsive to the public in Santa Monica as a result of the PEN system (Dutton et al., 1993).

Access to Information Technology. Equity is one of the most central issues raised by the critics of electronic service delivery. Many fear that electronic access will widen disparities between the information rich and the information poor, since the less well-to-do are unlikely to have the income, slack resources, or exposure to computing at work that will facilitate their access to and use of information technology as a medium for communication.

In Santa Monica, this issue was raised early in consideration of the launching of PEN and led the city to establish a number of public computer terminals from which individuals could access the PEN system. Even though nearly one third of Santa Monica households own a personal computer, almost one fifth (19%) of all accesses to the PEN system have been from one of the public terminals located at libraries and other public sites within the city (Dutton et al., 1993). The Department of Information Services has even had to develop rules, based on those established for sharing tennis courts, to govern how long an individual can stay at a terminal when others are waiting to use it.

In many distressed areas of the United States, particularly areas of the inner city, there is a very basic need to expose children and their parents to the existence and utility of information technology. The technology is simply invisible to many families. One of the primary reasons that households adopt a personal computer is to learn more about the technology per se (Dutton, Rogers, & Jun, 1987). Mere exposure to information technology is informative, irrespective of the use of the technology as a medium for accessing other information (Dutton, 1992a).

There are a number of identifiable needs for access to information technology within distressed areas. Interviews with public and voluntary organizations in distressed areas surfaced a need for systems to support scheduling and coordination among a variety of specialized agencies, such as central city schools. Currently, they lack good channels for communicating with one another on a day-to-day basis (Dutton, 1992a). One principal of an inner-city magnet school in Los Angeles argued that her school was electronically isolated in the sense that she lacked adequate facilities to schedule and coordinate activities with other nearby and suburban schools, resulting in a failure to take advantage of events, space, and other resources that could be more efficiently shared. The plain old telephone network was not sufficient to meet her needs and, therefore, she perceived a need for an electronic system that could support more asynchronous and efficient scheduling and coordination, such as an electronic bulletin board.

Access to technology is also perceived to be a need of minority business firms, which often lag behind more established firms in their use of information technology. This gap places the minority firm at a disadvantage, such as in competition to become primary or major subcontractors on projects. Reasons why they lack the technology are in part educational but also financial—they do not have adequate funds to purchase state-of-the-practice computers and telecommunication equipment. They must wait for systems to trickle down to their price ranges for them to be affordable. However, that situation leaves them technologically behind their major business competitors, which are using state-of-the-practice systems and who can take advantage of breakthrough technologies when they arrive.

Technology is not simply equipment, but also the know-how and expertise involved in using the technology. In this respect, there is a clear need for greater access to expertise and technical assistance in computing and telecommunications, particularly on the part of many voluntary, not-for-profit, and minority business firms. They often lack the technical consultants within their ranks, which are widely available to established business firms, who can advise and assist personnel with computing and telecommunication problems. Even if an individual can use a personal computer with some ease, that person might be brought to a standstill by a problem logging onto a online network, setting up their "autoexec" file, or any other of a countless number of minor challenges. One not-for-profit agency, CompuMentor, has been organized primarily to provide on-site assistance with computing to voluntary and not-for-profit enterprises, but the gap remains.

Public Information Needs Are Simple and Shared by a Large Proportion of the Public

Many cling to early visions of a mass market for the new electronic information services within the near future. However, most new media businesses have focused on niche markets. For example, while newspapers and other commercial service providers might eventually succeed in offering information to a sizeable proportion of the public, most have focused on electronic services that complement the newspaper, which they continue to view as central to their future. In the fall of 1992, *The Palm Beach Post,* owned by Cox Enterprises, began offering some weather, sports, and other information over an "easy-to-remember" three-digit telephone number. If such services can be offered at an incremental cost to newspapers and their subscribers, they might well become profitable even if utilized by a small proportion of the public.

As discussed previously, most of the successful information and communication projects in the public and private sector are oriented to serving information needs that are far more specialized, complex, and personal than often suggested by discussions of electronic publishing or for that matter the public sector equivalent of the ATM.[8] Contrary to the problems facing the banks and financial institutions, the problems of creating a public information service are not those of designing 20 screens of information that will be accessed billions of times. Quite the contrary. Public information needs are so personal that it is more likely to be a problem of designing thousands of screens of information that are uniquely tailored to a specific individual with a unique set of needs at a particular time and place. Some of the most successful innovations are responsive to nonroutine and specific questions or needs that are not shared by a large proportion of the public. Examples abound.

On Santa Monica's PEN system, there are literally hundreds of on-going discussions underway, many of which address the particular interests of only a handful of individuals. The French Minitel terminals provide access to over 15,000 different services. SeniorNet is focused on the needs of elders, particularly health care information (Arlen, 1991). Community Link, a joint effort of The Center for Community Change and Apple Computer, Inc., seeks to serve organizations that are focused on the economic development of low- and moderate-income communities. The network provides information on such matters as trust funds, financing, community development block grants, and subsidized housing, as well as a channel (electronic mail) for community advocates to communicate with

[8] It is clear that many within the computer industry view the ATM as a model for the public kiosk and the banking industry as the analogy to public sector agencies (e.g., Hanson, 1992).

one another about their common problems. HandsNet, another nonprofit information and communications network, is also specialized, focusing on serving agencies providing human services.

The specialized information needs of individuals are also evident in the activities of information and referral agencies. INFO LINE keeps records of the requests for information it receives. Over 1 year, the requests from more than 200,000 calls fell into nearly 500 categories. The only categories that might be called routine were the three most frequent requests, which were for emergency food, emergency shelter, and assistance with restoring gas, electric, or other utilities. An index of public information needs developed by the INFO LINE staff took years to develop and fills a 3-inch-thick directory that looks like the telephone book for a major metropolitan area.

The proliferation of audiotext "900" numbers is another reflection of the diversity of individuals' information needs. Specialized services, such as *USA Today*'s "The College Info-Line," that can be of value to only a small proportion of the public but on a national or regional scale, seem especially well suited for this type of specialized service.

People Want Information at Their Fingertips

Early visions of the public information utility were based on the assumption that the public is interested in information, but this assumption might well be a misleading guide to the development of services. In fact, many assume that the public is composed largely of avid information seekers. However, the experiments with electronic services we surveyed tend to reinforce an observation made by many involved in the early market trials of videotext services, which is that the public might be more interested in specialized services and in communication than in information per se. As Hooper (1985) said in discussing the lessons learned in the marketing of Prestel, Britain's innovative videotext service, "Prestel was invented and launched on the assumption that there is a large and ready market for electronic information retrieval—retrieving pages of information stored in computers. Reality turned out differently" (p. 190).

This point is striking from our research on the Santa Monica PEN system. PEN provides a menu of services, which includes several options for electronic access to static information about the city, the council, and community events—a read-only menu. PEN's "Mailroom" provides the capability for electronic mail. From the outset of the PEN system, conferences generated more interest, as gauged by the number of accesses to this facility, than did the bulletin boards. Moreover, there is a trend toward an increasing focus on conferences and electronic mail as opposed to simply retrieving information stored online (Dutton et al., 1993).

This phenomenon is not new. The ARPANET was originally developed to support remote access to computing services, but evolved instead to become primarily a medium for interpersonal communications. SeniorNet, HandsNet, Community Link, Cleveland's FreeNet and other computer-based information services for the public have made electronic mail and conferencing a central aspect of their operations.[9]

In fact, as we found in observing a number of agencies that are expressly focused on providing "information" to the public, the boundary between information and communication can be quite problematic. Even agencies that are explicitly devoted to the provision of information services can find themselves offering a communication service. For example, information and referral agencies, such as INFO LINE, are called by individuals needing particular information, whether it be food, shelter, counseling, financial assistance, or other services. Even the most casual observer of the information and referral function could not help but see that these calls normally generate an extended conversation about the nature and history of one's situation and a set of recommendations regarding the options and problems facing the individual. The information specialists provide counseling and emotional support of another human being—not just information.

The distinction between communication versus information services is not simply a semantic issue because it has significant implications for policy and practice. Instead of focusing attention on providing access to specific information content, which is a major theme of many scholars interested in public access, we should focus more attention on access to information technology as an emerging medium for communication. It is access to the technology as a medium for communication versus specific informational content that is paramount. Similarly, we should view the public less as an audience or even as users of information services, but as providers of content and services. This realization could also change the way we look at the information service agencies themselves, since they might be better conceived of as networkers or communication firms than as content providers or publishers.

Of course, the reality is that both perspectives are true with the new electronic services. The public is both audience and provider; the public needs access to information as well as to information technology; and information providers are publishers but also carriers or transmitters of information provided by others. The point is that the significance of the communication functions served by the new media is too easily and inappropriately marginalized in discussions of access to public "information" unless information is defined in its broadest sense.

---

[9]An excellent discussion and overview of many of these online services is provided by Aumente (1989).

It Doesn't Matter Where Information is Stored

One technically rational but dubious assumption of the age of electronic information networks is that it no longer makes any difference where information might be physically housed. If you work or sit at a computer, so this argument goes, modern telecommunications enable you to work as easily from the opposite coast as from the same office in which the computer is located.

One theme emerging from a variety of relatively successful public and not-for-profit information systems is that systems benefit greatly from a psychological sense of community ownership. SeniorNet's success is in part a function of its perceived ownership and control by seniors. Users feel a part of a community, rather than simply the market of a new media entrepreneur (Arlen, 1991). PEN users are PENners and residents of Santa Monica—literally members of a community. Community leaders in the inner city of Los Angeles do not want their communities viewed as a new market for information technology; they want to gain a sense of ownership and control over information technology that they design to serve their own community's needs (Dutton, 1992a).

It might well be that the development of any national information system must always balance efforts to achieve economies of scale and scope with efforts to develop and maintain a sense of community and ownership. Those who are expected to benefit from the use of particular services need to truly feel that they are being served by a system that is controlled by their community. Their community might be defined by categories of needs and interests, such as senior citizens, or by geography, as in the case of Santa Monica residents.

Implications for National Policy

The appropriate national policy is more complex than suggested by much of the discussion of "a national information highway." The federal government might undermine the development of electronic service delivery by promoting any one medium for reaching the public with information perceived to be in the general public interest. If the emerging trends discussed here are any guide to the future, federal strategies for supporting the public's information and communication needs should focus on facilitating the migration all sorts of specialized information and services to all kinds of electronic media, including broadcasting and telecommunications as well as newer information technologies like voice processing, computer bulletin boards, and multimedia kiosks.[10] Federal, state, and local agencies

[10] A perspective on federal policies aimed at facilitating the development of electronic service

need to support national initiatives to coordinate the diffusion of public, not-for-profit, and private electronic networks and services throughout the United States.

At the same time, the public and not-for-profit providers of information services must insure that the public has the capability to access electronic information. This access will come through the provision of public facilities and support to groups that will be electronically disadvantaged, if not disenfranchised, without "on ramps" to the many information highways crisscrossing the United States.

There is also a federal role in tying together the growing multiplicity of electronic networks and services created at the federal, state, and local level. It can do this by setting standards, interlinking networks, and providing an indexing function so that the physical location of information or agencies is increasingly irrelevant to the citizen needing access to a particular service. In these ways, innovations in the provision of electronic services by public and not-for-profit agencies at the local level can be complemented by the development of regional, state, and federal initiatives to construct a seamless national network of networks for the public as opposed to an electronic maze of standalone applications.

Political jurisdictions, along with not-for-profit and other quasigovernmental agencies, can become valuable providers of electronic services that the public can access from their homes and public facilities over a variety of media. However, their efforts will be limited if not diminished if Federal policies do not create incentives and mechanisms to provide the on ramps, off ramps, and highways to make it more universal and equitable to the U.S. public. But it will not be neatly integrated on a single, national, public electronic network. Instead, it will evolve bit by bit as specialized systems developed by early innovators, such as those discussed in this chapter, diffuse and become linked into larger regional and statewide systems of public, private, and not-for-profit networks that support broadcasting, transactions, access to public information and records, and interpersonal communications over broadcast, cable, and telecommunication networks. The wired nation will be a complex and decentralized mosaic of specialized networks rather than a single, federally planned and integrated information highway.

SUMMARY AND DISCUSSION

The primary rationale behind promotion of the public information utility in the 1960s, like that behind videotext in the 1980s, was the public's interest

---

delivery is provided by this book as well as by a forthcoming report by the Information and Communication Technology Program of the OTA, entitled "Making Government Work."

in getting information. The market failure of interactive cable television and, later, of commercial offerings of videotext services undermined this rationale to a great extent but not entirely. Many developers of information systems continue to believe that a substantial proportion of the public is actively seeking information and will use new information systems if the right information is packaged in appropriate ways. This is one line of reasoning behind the growing interest in multimedia personal computers and the Internet as well as new telephone services, such as the screen phone, for the delivery of electronic information services.

However, the public and not-for-profit sectors are driven also by a variety of other rationales to develop mechanisms for electronic service delivery. Some distressed segments of the public lack basic information of relevance to their health and safety. Among politically active citizenry, such as in Santa Monica, a growing number expect more direct and convenient ways to participate in public affairs and communicate with public officials and agencies. Public as well as not-for-profit agencies share an interest in reducing the costs of providing routine information to the public and stand to benefit from systems that will support communication within communities in which traditional forms of interpersonal communication are perceived increasingly to be insufficient. The growing prominence of ethnolinguistic communities demands that public agencies be more responsive to the cultural diversity of their clientele.

The success of the public and not-for-profit sector in using electronic media will create a need to address issues over the equity of service provision across jurisdictions and socioeconomic groups. Equity considerations sometimes moderate interest in cutting edge technologies, since their expense is likely to reinforce socioeconomic disparities, but these same equity concerns can support the extension of more universal access to electronic communication and information technologies that are already in use (Williams & Hadden, in press).

Despite a number of genuine needs for extending electronic information services, these experiments face formidable problems. One is that the information needs of the public are complex, multifaceted, and unique to individuals, making it difficult and costly to adapt systems, such as an information kiosk, that are more easily used to provide programmed responses to routine inquiries, such as a bank balance. Another is that information resources are often so fragmented across jurisdictions and organizations that they do not match well with the needs of individuals. Information that is relevant to these needs is exceedingly difficult and costly to create and maintain in a way that is comprehensive, accurate, and up-to-date, resulting in only a few truly valuable information systems that serve the public interest.

Electronic information services are raising problems of their own, including concerns over the equity of electronic service provision and the rules governing the privacy, confidentiality, and freedom of expression over electronic media. If a patchwork of isolated electronic service delivery systems emerges, they might well exacerbate rather than diminish inequalities in access to information resources, not just to the poor, but also to any member of the public who faces educational, language, or physical handicaps that create a barrier to accessing information or services.[11]

Electronic service delivery is not a quick technological fix to fundamental problems of citizen access, which often extend from the growing diversity of urban centers, the financial crises facing U.S. governments, and the basic needs of the poor and unemployed. However, failure to develop the infrastructures and applications to provide all sorts of electronic services to the public might well worsen these problems within an information society in which all kinds of services are increasingly being mediated by broadcasting, telecommunication, and other electronic media. Information technology can be used to facilitate citizen access, to make agencies more responsive to the public, and to better meet the needs of many specialized publics for information about health and social services.

REFERENCES

Arlen, G. (1991). SeniorNet services: Toward a new electronic environment for seniors. In *Forum Report* (Vol. 15). Queenstown, MD: The Aspen Institute.

Aumente, J. (1989). Online for social benefit. In *Forum Report* (Vol. 12). Queenstown, MD: The Aspen Institute.

Davidge, C. (1987) America's talk-back television experiment: QUBE. In W. Dutton, J. Blumler, & K. Kraemer (Eds.), *Wired cities: Shaping the future of communications* (pp. 75–101). Boston: G.K. Hall.

Dutton, W. H. (1992a, September 15). *Electronic service delivery and the inner city.* Community workshop summary: Overview of a community workshop held for the Office of Technology Assessment U.S. Congress at The Annenberg School for Communication, University of Southern California, University Park, Los Angeles.

Dutton, W. H. (1992b). Political science research on teledemocracy. *Social Science Computer Review, 10*(4), 505–523.

Dutton, W. H. (1992c). The social impact of emerging telephone services. *Telecommunications Policy, 16*(5), 377–387.

Dutton, W. H., Blumler, J. G., & Kraemer, K. L. (Eds.). (1987). *Wired cities.* Boston: G.K. Hall.

---

[11]Many efforts to support electronic service delivery for the average person have generated benefits for individuals with handicaps. For example, electronic bulletin board systems have been an excellent means for individuals with hearing impairments to communicate with others on a more equal basis. However, some electronic information services, such as ATMs and their public kiosk equivalents, could limit access by individuals with impaired eysight, unless they are specifically designed to support them.

Dutton, W. H., & Guthrie, K. (1991). An ecology of games: The political construction of Santa Monica's Public Electronic Network. *Informatization and the Public Sector, 1*(4), 1-24.

Dutton, W. H., Guthrie, K., O'Connell, J., & Wyer, J. (1991). *State and local government innovations in electronic services* (Unpublished report for the Office of Technology Assessment, U.S. Congress). Los Angeles: Annenberg School for Communication, University of Southern California.

Dutton, W. H., Rogers, E. M., & Jun, S. (1987). Diffusion and social impacts of personal computers. *Communication Research, 14*(2), 219-250.

Dutton, W. H., Wyer, Wyer, J., & O'Connell, J. (1993). The governmental impacts of information technology: A case study of Santa Monica's Public Electronic Network. In R. Banker, B. A. Mahmood, & R. Kauffman (Eds.), *Perspectives on the strategic and economic value of information technology investment* (pp. 265-296). Harrisburg, PA: Idea Publishing Group.

Elton, M., & Carey, J. (1984). Teletext for public information: Laboratory and field studies. In J. Johnston (Ed.), *Evaluating the new information technologies* (pp. 23-41). San Francisco, CA: Jossey-Bass.

Fountain, J. E., Kaboolian, L., & Kelman, S. (1992, October). *Service to the citizen: The use of 800 numbers in government.* Paper prepared to the Association for Public Policy and Management, Denver, CO.

Forester, T. (1989). *Computers in human context.* Cambridge, MA: MIT Press.

Guthrie, K., Schmitz, J. Ryu, D., Harris J., Rogers, E., & Dutton, W. (1990, September). *Communication technology and democratic participation: PENers in Santa Monica.* Paper presented at the Association of Computer Machinery's (ACM) Conference on Computers and the Quality of Life, Washington, DC.

Hanson, W. (1992, September). The kiosk phenomenon, *Government Technology,* pp. 16-17.

Hooper, R. (1985). Lessons from overseas: The British experience. In M. Greenberger (Ed.), *Electronic publishing plus* (pp. 181-200). White Plains, NY: Knowledge Industry Publications.

Lucky, R. W. (1989). *Silicon dreams.* New York: St. Martin's Press.

Noll, A. M. (1980, March). Teletext and videotex in North America. *Telecommunications Policy,* pp. 17-24.

Noll, A. M. (1985). Videotex: Anatomy of a failure, *Information and Management, 9,* 99-109.

Noll, A. M. (1992, October 27). *A national information system.* Paper presented to the Freedom Forum Conference, Columbia University, New York.

Pool, I., & Alexander, H. E. (1973). Politics in a wired nation. In I. de Sola Pool (Ed.), *Talking back: Citizen feedback and cable technology* (pp. 64-102). Cambridge, MA: MIT Press.

Sackman, H., & Boehm, B. (1972). *Planning community information utilities.* Montvale, NJ: AFIPS Press.

Sackman, H., & Nie, N. (Eds.). (1970). *The information utility and social choice.* Montvale, NJ: AFIPS Press.

Westen, T., & Givens, B. (1989). *The California channel: A new public affairs television network for the state.* Los Angeles, CA: Center for Responsive Government.

Williams, F. (1991). Network information services as a new public medium. *Media Studies Journal, 5*(4), 137-151.

Williams, F., & Hadden, S. (in press). On the prospects for redefining universal service.

CHAPTER 13
Creating an Electronic Information Services Marketplace in the United States

Charles Steinfield
Michigan State University

As early as 1974, visions of an interactive electronic information marketplace in the United States were stimulated by the emergence of techniques for linking communications networks to computers (see Baer & Greenberger, 1987). Even before the personal computer industry appeared and began targeting the residential market, various players from the broadcasting, cable, telecommunications, publishing, computer, and computer services industries flirted with the idea of delivering electronic information services to the mass market (Steinfield, 1992). Some players were heavily influenced by developments in Europe, where one Post, Telephone, and Telegraph (PTT) administration after another sought to establish national videotex and/or teletext services, mostly with little success (Bouwman, Christoffersen & Ohlin 1992).[1] There was particular interest in the French experience, where a unique approach to the mass market did result in what many believe to be the world's most successful videotex system (Thomas, Vedel, & Schneider, 1992). However, the U.S. telecommunications and information services environment is quite different from other countries, and, as a consequence, the electronic information services industry has evolved quite differently here. In this chapter, I briefly describe the development of the electronic information services marketplace, contrasting it with the French experience. Research on the use of electronic information services in France and the United States, particularly by business users is then described.

[1]In Europe, until only very recently, nearly all telecommunications services were provided by government-owned monopolies known as PTTs. Sometimes branches of government, sometimes government-owned companies, the PTTs saw the provision of electronic information over telephone networks as a means of increasing revenues derived from the residential market. Beginning with the Prestel (originally called Viewdata) service in the United Kingdom, most of these services offered alphanumeric information enhanced by crude graphics, in a page-oriented format. Television sets equipped with modems and decoders were the display devices. Simple menu-driven control enabled people with no computer training to access the various services available. With the exception of the French Teletel system, all of the European videotex trials did not achieve significant market penetration.

These contrasts then permit us to draw several important lessons that can serve as guideposts for policymakers as they shape the emerging electronic information marketplace in the United States today.

THE EMERGENCE OF A COMPETITIVE DATA NETWORK INDUSTRY IN THE UNITED STATES

To understand the U.S. electronic information services industry requires some basic background knowledge about the policies and institutional structure of telecommunications. In the past, the U.S. telecommunications industry was always comprised of private monopolies operating under various regulatory controls. For most of this century, the Bell system under AT&T controlled some 80% of local telephone service and until deregulation in the 1970s, nearly all long distance service as well. The system of private, regulated monopoly was justified on the grounds that telephone services were "natural monopolies," where optimum efficiency and lowest cost for provision of services were best obtained by a single network provider in any given area. In other countries where government provision of services was more acceptable, the model of a single national monopoly for telecommunications administered by the government was far more common. In the United States, although basic telephone service was regulated, other related industries were not. Traditionally, for example, we have considered the computer industry to be a competitive industry, not subject to the same conditions that led telephony to be a natural monopoly. In addition, the provision of information—through newspapers, books, magazines, and even the radio and television broadcast media, were likewise considered competitive. These industries were not subject to the same kinds of regulation as telephony except for licensing practices made necessary by the need to manage the radio spectrum in an efficient and fair manner.

This brief background should help, then, in understanding the far-reaching consequences of a decision made by AT&T back in 1956, in the process of settling an antitrust suit brought against the company in 1949. Essentially, we might say that the Justice Department was only a reluctant party to the natural monopoly argument, and most strongly disagreed with the vertical integration in AT&T that extended all the way to the provision of all of the equipment—from telephones to switching equipment. In that consent decree, AT&T agreed to stay out of all unregulated businesses, in return for keeping their equipment division. In particular, AT&T was to stay out of the computer and computer communication business, which at the time was relatively insignificant. No one foresaw the dramatic growth of computers, nor the accompanying future demand for the transmission of data between them.

As the demand for data communications networks grew, policymakers were faced with the challenge of determining whether these types of services were "basic," and hence could be offered by AT&T as a regulated monopoly service, or "enhanced" and best provided under a competitive policy regime. In the 1970s, policymakers generally agreed that a new type of network optimized for data communications—known as a packet-switched network—was an enhanced service. AT&T was welcome to provide the underlying basic transmission, but faced constraints resulting from the earlier 1956 Consent Decree when it came to offering the enhanced services that packet-switched networks provided. Such services as error checking, messaging services, protocol conversion, and so forth, were all deemed to be enhanced, and therefore, competitive (for reviews of U.S. policy in this area, see Matos, 1988). Although some mechanisms were sought whereby AT&T and its Bell Operating Companies could enter the packet-switched network market, especially in a series of FCC proceedings known as the Computer Inquiries, the main effect of policies at the time was that other entities besides the traditional Bell System companies dominated in data communications.

Although many of the restrictions on AT&T's entry into the computer and computer communications business were ultimately removed at the time of the AT&T divestiture in 1984, restraints on the provision of electronic information services remained to some extent. AT&T faced a 7-year moratorium on this activity, and the divested Regional Bell Operating Companies (RBOCs) also were restricted from providing any information services. This line of business restriction was only removed by court order in late 1991.

On the one hand, we can consider the policies that restrained AT&T from dominating the data communications market a great success. The U.S. data communications market is considered one of the world's most innovative. A number of firms offered packet-switched networks, including such early entrants as Telenet and Tymnet. Many firms interested in offering online services benefited from either having their own network, such as CompuServe or from having lower cost network providers through which their services could be made available. In contrast, information service providers in other countries often could only use the national monopoly provided packet-switched network. The higher rates and less flexible usage policies imposed by PTTs made it difficult for information service providers to compete effectively. The local area network industry also flourished first in the United States, developing approaches that were completely distinct from the logic of the switched telephone network. And finally, the uncontrolled development of computer communications, coupled with liberal usage policies on private leased lines, opened the door to the most phenomenal data network of all, the Internet. Not a single network, but rather a network of networks, this complex consortium of data networks is provided by a combination of government support and grass roots participation. It is unlikely that such an entity would have emerged if we had a single, monopoly provided data communications infrastructure.

On the other hand, the policies that encouraged a highly competitive data network industry in the United States were not without their downside. It can be argued that in the absence of any form of central oversight, the United States ended up with a fragmented data network infrastructure characterized by a large number of separate, proprietary networks having little interconnection or interoperability between them. Subscribers to Telenet could not seamlessly communicate with subscribers to Tymnet. Customers of one online service were not guaranteed access to information services residing on another network. Network providers concentrated their efforts on the more lucrative business market, resulting in negligible participation by households, small businesses, K–12 schools, and local governments.

How then, has this situation influenced the development of an electronic information services in the United States? Essentially, it has led to a dynamic, innovative, but clearly niche-oriented online industry, that caters more to the business than residential market. There is not yet an electronic information services "mass market" in the United States. Several noteworthy attempts to provide easy to use videotex systems to the residential market ended in failure in the mid-1980s (see Steinfield, 1992). Systems that catered to the home computer market did experience more success, but these still reach a relatively small percentage of households. The vast majority of the revenue generated by electronic information services comes from financial services offered to the business market (Steinfield, 1992). The limited telephone company participation in the electronic information services industry has contributed to a general fragmentation of the market, limited geographic scope and scale of the systems that have been introduced, and lack of integration with existing telecommunications services that might have encouraged greater use.

VIDEOTEX IN FRANCE

We can now contrast the U.S. situation with that of France, to show how an electronic information services market evolved under a totally different policy and institutional structure. Although we make no attempt to argue that the French model be copied here—such an approach is both impractical and infeasible (see Kramer, 1991)—there are nevertheless some important lessons that we can learn about the nature of mass market information services.

In 1983, after several trials, the French PTT (now called France Telecom, but known then as the Direction Generale des Télécommunications or DGT), commercially launched the Teletel videotex system. At the time, the DGT was a government administration with a total monopoly over all telecommunications services and network infrastructure.[2] They could thus introduce videotex on a national basis, free of any competitive pressures. But many other national videotex systems were started under the same monopoly conditions, so this alone does not explain Teletel's success. For example, both the British Prestel system and the Bildschirmtext system in Germany were introduced around the same time as national systems by the then monopoly telecommunications entity, but did not achieve widespread penetration (Thomas et al., 1992).

We can point to a number of widely recognized strategies followed by France Telecom that made their experience with videotex different. Perhaps most often discussed is their massive effort to "create" a market for information services by giving away the Minitel terminals that telephone subscribers used to connect to the network. At one point, more than 1 million terminals were distributed per year, and today more than 6 million have been distributed. In order to offer terminals free of charge, they essentially contracted with equipment suppliers for millions of Minitels, thus ensuring that the terminal costs were lower due to economies of scale in production. Only the basic model was given away; more sophisticated Minitels were rented. This strategy ensured that there would be a population of users capable of accessing services.

A second strategy was to develop an "anchor" application that gave subscribers reason to access Minitel in the first place. This was the electronic *annuaire téléfonique* or electronic directory. This directory was a searchable, national listing of both residential and business subscribers. Moreover, businesses could "advertise" with extra pages displaying more information about their products, services, locations, business hours, fax numbers, and so on. France Telecom even tried to limit the production of printed directories in order to require use of Teletel, but public pressure forced the continued distribution of directories to those who requested them.

A third strategy was to proactively encourage information service providers, while following policies that made becoming such a provider as easy as possible. Thus, France Telecom worked with several prominent compa-

[2]The DGT is now a state-owned company, and no longer has a monopoly on value-added services. The monopoly on basic services is due to end in 1998 and there are even discussions now to end the monopoly on the provision of telecommunications infrastructure. It is thus unlikely that the method of introducing the Teletel system could, in fact, be replicated today, even in France.

nies to help them establish services, even underwriting some of the service development costs. Subsidiary firms of France Telecom maintained videotex hosts for companies that could not afford their own and provided consulting on marketing and software development. They even defused a potential confrontation with the French newspaper and publishing industry by giving major publishers privileged access to Teletel. They followed a common carrier approach to the provision of services, deciding not to enter into any information service business themselves that would be in direct competition with third-party service providers. And users benefited from the fact that all services are available on the same network using the same access numbers.

A fourth strategy was to strictly enforce technical standards for service providers. Not only did this ensure that all services worked when accessed, but the "look and feel" of services was consistent.

A fifth strategy was to use their own national packet-switched network (Transpac) to carry videotex traffic. This made it easy to achieve national coverage, and meant that information providers (IPs) could connect to the network from virtually anywhere in the country without incurring any transport cost penalties. IPs thus managed their own databases, making it easier to ensure timely updating. The extra traffic generated by videotex, while creating some capacity problems at first, helped to justify investment in and create revenue for Transpac.

A sixth strategy was the now famous kiosque tariffing system, whereby users call a single number to access a large set of services, all of which are charged at the same rate. In fact, at the start there were only four numbers to dial: 11 for the electronic directory, 3,613 to reach services that were free to the caller (like an 800 number), 3,614 to reach services in which there was a communications charge but no service provider charge, and 3,615 for the bulk of the services that included a communications charge and per minute charges. This simplified, "visible" pricing system means that all users know the costs by the number they dial, and these numbers are the same everywhere in the country. No subscription is necessary, allowing spontaneous and immediate access when desired. All billing and collection is done by France Telecom, which allows much smaller information service providers to participate. Today there are several additional numbers for higher "value-added" services, but the logic of the kiosque remains the same.

The results of France Telecom's approach are impressive, the debate over whether they have actually realized a profit notwithstanding.[3] The more

[3]One audit by the French government was critical of the Teletel policy, arguing that it had resulted in substantial losses. Their criticisms have themselves generated debate over how to properly determine the relative profit or loss associated with Teletel, including questions of whether such positive consequences as the extra revenue videotex generates for Transpac, the

than 6 million terminals are in use in nearly 100% of French firms, and more than one quarter of French households. They can also be found in every post office, in addition to other public places. Nearly everyone in France has access to the Teletel system either at home or at work. There are now over 20,000 service codes on Teletel, and the number of service providers continues to grow.[4] Many of the services that we often refer to as a future benefit of interactive networks are taken for granted in France. People routinely access the schedules of all forms of public transportation by Minitel, and on trains and airplanes can even make reservations. Hotel rooms can be booked, merchandise from catalogues can be ordered, bank accounts can be managed, and a variety of home shopping can be accomplished. There are many services offering both general news and very specialized information (such as where in Paris one might find a specific movie, including the time of showing, price, size of theater, critics' reactions, etc.). Many government services are available electronically, and the national telephone directory allows people to search for numbers themselves, even when they don't know first names, exact addresses, and the like.

BUSINESS IMPLICATIONS OF A MASS MARKET IN ELECTRONIC INFORMATION SERVICES

Although most of the publicity surrounding videotex in both France and the United States has focused on the residential mass market, these services can have significant implications for the business community as well. Organizations have long attempted to create and maintain competitive advantage through information and telecommunications networks (Keen, 1988; Porter & Millar, 1985). These networks have increasingly been extended up and down the "value chain" to include links to suppliers, distributors, retailers, and ultimately to customers (Cash & Konzynski, 1985). Mass market connections are not necessarily a part of this process in the United States, however. Rather, such applications as "electronic data interchange" between closed groups of suppliers and manufacturers on private networks are

value to the economy of the information services industry, and the raising of the general level of computer literacy of the French population should be counted (Bouwman & Latzer, 1994). A recent independent assessment of Teletel by Coopers and Lybrand estimated that the total investment will be recouped by the late 1990s.

[4]Here, too, considerable debate has raged over the "quality" of services, since early growth was fueled by chat services known as *messageries* and the so-called "pink" or erotic Minitel services. Today, however, these types of services account for a small percentage of use and they have largely been replaced by more mainstream services.

flourishing, despite the limited end consumer participation in data networks (Steinfield, Kraut, & Streeter, 1993).

In France, the presence of a ubiquitous and open data network has impacted on interfirm business data networking. A series of case studies of business use of the Teletel system shows the utility of the Teletel system as a platform for interfirm electronic transactions despite its rather limited transmission capacity (Steinfield & Caby, 1993; Steinfield, Caby, & Vialle, 1992). Most transactions do not actually require significant bandwidth, and include such generic applications as order processing and reservations systems. Using a data network for these applications conveys significant cost advantages, including lower costs for orders due to labor savings and fewer errors. Furthermore, because information is in computer-readable form right from the start, firms are able to exploit the information by-products of transactions as a competitive weapon. For example, transaction summaries such as purchase histories can be valuable information to feedback to buying firms, helping them to manage their own procurement practices.

Because of the ubiquitous access to Minitels, however, important differences between the use of data networks in France versus the United States have begun to emerge. Many firms in France take advantage of the large numbers of "end customers" reachable by a data network to offer revenue generating services. Because a switched public network is used for connections, even the smallest users can participate. Hybrid networks appear, with leased lines to users that generate large volumes of traffic, but with Teletel access for the residential and small business customer (Steinfield et al., 1992). Moreover, mass market access has opened the door to another kind of business, that of the "market maker" who establishes "electronic trading networks.[5] The LAMY company, for example, has created a spot market for the French trucking industry by posting real-time offers by freight forwarders. Independent truckers planning to haul freight between two cities can fill any excess capacity in their trucks by checking in with their Minitel. They call on the normal kiosque (generating revenue for each call to LAMY), input their city of origin, destination, and available capacity, and all matching offers are found. With 25,000 independent truckers calling on a more than daily basis, a totally new market has been created. This electronic market would not be possible without an open, public data network with ubiquitous connectivity.

There is some evidence that France's Teletel network also conveys some of the benefits of computer networks to smaller firms that would not otherwise be able to afford them. In a survey of both U.S. and French

[5]See Mansell and Jenkins (1992) for a discussion of the characteristics of electronic trading networks, including their strategic benefits and policy implications.

firms, the use of data networks for interfirm transactions was compared across the two countries (Streeter, Kraut, Lucas, & Caby, 1993). Across the general population of companies in both countries, large firms were far more likely than small firms to use networks to link with trading partners. However, among firms that used the Teletel network, there was no size difference at all, with small firms just as likely as large ones to engage in electronic transactions with trading partners.

These findings suggest that having a common, open data network conveys the same kinds of *network externalities* as a having a totally interconnected voice communications market does. Users benefit from other users because they help to constitute a *critical mass* that attracts suppliers. Service providers likewise benefit from the presence of other service providers because they help to attract more users, much like stores in a shopping mall or district attract more total buyers. And use of the public network permits the costs to be shared over larger numbers of users, lowering average costs and enabling the participation of small firms and households.

AUDIOTEX AS AN ELECTRONIC INFORMATION SERVICES MASS MARKET

Although the number of homes connected to consumer-oriented online services like Prodigy, America Online, and CompuServe continues to grow, and approaches 5 million in the mid-1990s (Arlen, 1994), this is still fewer than 6% of the nearly 92 million U.S. households.[6] This is not to say that other Americans have no means of electronically accessing information or engaging in transactions, however. Rather, the dramatic growth in the use of audio information services accessed by special dialing prefixes like "800" and "900" represents an alternative method of providing electronic information services. We have long been accustomed to calling an 800 number to speak with a customer service representative of a company with which we wish to do business. However, increasingly, calls to these numbers are first answered by an automated attendant, and the use of a technology called *interactive voice response* allows callers to press buttons on a touch-tone phone to step through menus until they reach desired information. When the calls are made using a "900" or "976" special access code, with the caller paying both a communications and an information services charge, we

[6]Arlen (1994) estimated that CompuServe and America Online have 1.8 million and 700,000 subscribers, respectively, with Prodigy connected to about 1.2 million homes (with multiple registered users per home). There are an estimated 92 million households in the United States (U.S. Dept. of Commerce, 1990).

generally refer to it as *audiotex* (see Steinfield & Kramer, 1994, for a review of the U.S. audiotex industry).

This relatively new industry exhibited extremely rapid growth throughout the late 1980s. Viewing the audiotex industry as an electronic information service helps to highlight some of the factors that contributed to its early success. First, access is through a widely available technology—the push-button telephone. Second, no specific training or skills are needed. Third, the pay-per-call nature of the service enables spontaneous access with no presubscription, just as with the French Teletel kiosque. Fourth, telephone companies offer billing and collection services as well as service bureaus so that information providers do not need their own interactive voice response equipment. Fifth, the common language of touch-tone signaling and audio response means that service providers can utilize any network provider, and still be reached by any subscriber regardless of what kind of equipment they have or who their local and long distance carriers are. Hence, there is both interconnection and interoperability. It is also worth noting that the long distance audiotex services fared much better than the local "976" services because "900" numbers gave information providers access to a national market (Steinfield & Kramer, 1994).

The audiotex industry is not without its controversies, and the ease of entry made it particularly susceptible to service providers engaging in consumer fraud or offering pornographic services (Glascock & LaRose, 1992; Samarajiva & Mukherjee, 1991; Steinfield & Kramer, 1994). These abuses caused a strong backlash by subscribers, telephone companies, and regulators, resulting in more stringent controls and a significant drop in revenues from the nearly $1 billion for the industry in 1990 (Steinfield & Kramer, 1994). Despite the problems associated with the "900" services industry, the use of "800" numbers by business is still flourishing. It represents a tremendous strategic tool for the business community, and currently is their only real electronic link to the mass market.

Although we can think of audio-based information services using the telephone as the home terminal as a type of electronic information service to the mass market, it does have significant input and output limitations. Telephones do not have keyboards for alphanumeric input or screens for text and graphic output. The services now in use generally translate the tones from a push-button phone into numbers reflecting menu choices. The "data" to be output is then either a prerecorded message or synthesized voice read from a database. Although there are hybrid forms of output, such as "fax-back" services, these are not generally directed at the home market. Because the menus are read aloud to callers, there is a real limit to the number of options and information that can be accessed using audio. Thus, the overall service complexity is far more limited than would be possible using keyboard input and screen output. We are thus still seeking a more robust electronic information service for the U.S. mass market.

THE INTERNET AS AN ELECTRONIC INFORMATION SERVICE

No discussion of electronic information services can ignore the recent phenomenal growth of the Internet, which by the mid-1990s interconnects some 20 million users ("Government/Industry Panel," 1994). Many observers in industry and government now believe it represents the most important model for the provision of electronic information services. There is not enough space to give the subject of the Internet full treatment here (see Quartermann, 1994, for a good history of the Internet). Presently, most current users are affiliated with research organizations and universities, although many K–12-level schools, public libraries, and small and large businesses are rushing to connect to the Internet. The Internet now is being flooded with commercial services in much the same way that information service providers flocked to Teletel once there was a large enough user community. Although it does not yet connect to the home consumer market to any meaningful degree, this is likely to occur quickly. All consumer online services have announced plans for Internet connectivity.

The reasons behind the growth of Internet are similar to the Teletel experience. Information service provision is completely decentralized, in fact even more so than on Teletel because every user is encouraged to be an information provider. Proprietary protocols are not used; the emphasis is on open protocols conforming to a minimum standard. A critical mass of users has come mainly due to some initial subsidization of access costs, by the institutions that collectively comprise the Internet, and by the government. The final problem to be solved to enable as Internet use by the general population, was to simplify the user interface, much as Teletel designers sought successfully to do. This occurred with the creation of the World Wide Web and Mosaic (and similar programs). The World Wide Web interconnects servers connected to the Internet, and makes it easy for users to find information through the use of hyperlinks. Hyperlinks essentially allow users to select a keyword, picture, or phrase in order to connect to another file or computer where additional desired information resides. Users do not need to be concerned with the actual process of establishing computer connections, and desired data may actually be stored on a computer halfway around the world. Mosaic and similar *browsing* programs integrate the ability to use the Web and other Internet functions in an easy to use program. Like the Minitels, these programs have been distributed widely and freely to create an immediate mass market. Millions of copies have been downloaded by users everywhere, and in less than 2 years, the Internet, the World Wide Web, and Mosaic have become household words.

IMPLICATIONS FOR THE FUTURE OF ELECTRONIC INFORMATION SERVICES

This review of both U.S. and French experiences with electronic information services to date illustrates many important issues for policymakers, network operators, and service providers. The almost nonstop flood of announcements of new trials for the provision of interactive services by cable television companies and the newly unshackled RBOCs should give us pause as we reflect on past experiences and current policies. In this section, I summarize the lessons that can be drawn from the review.

One lesson to be drawn is the value of *universal access* in establishing an information services mass market. Niche services marketed directly to a very specific market segment can be profitable, but are not likely to serve as platforms for the kinds of innovative service creation witnessed in France on Teletel. In addition, the attempts that were confined to small localities also fail to generate the kinds of network externalities necessary to attract a critical mass of users and service providers. Experience, along with a good deal of theory, has shown that the value to potential users of an interactive network grows with the size of the community connected (Markus, 1990). This implies that strategies to connect households, small businesses, and schools to an electronic information service are needed. Such strategies might include versions of services that can be used with lower cost terminals, charging mechanisms that permit those with limited means to still have some form of access, and perhaps targeted subsidies to facilitate access.

Achieving universal access in the U.S. context will not be easy. The abundance of private networks and proprietary protocols, as well as the increasing fragmentation of the public network in both long distance and local exchange service create problems of interconnection and interoperability. In terms of interconnection, we need to ensure that connection to one network operator does not preclude access to the users of another network operator. However, policymakers as well as network operators must also be vigilant to ensure interoperability across network boundaries. If past experience is any guide, this is likely to be one of the most difficult problems to solve. Witness how long it took, for example, for a person who subscribes to one cellular telephone company to be able to use his or her cellular phone in another network operator's region. The technical constraints on "roaming" were not really the issue—it was the business arrangements between companies that had to be satisfied. Now add the complexity of many more companies offering network infrastructure, information that is not just voice or text, but multimedia, and the need for high level applications software, and the challenge of making sure that services function transparently across multiple network operators becomes even greater.[7]

Another lesson from Teletel that is likely to be unpalatable for American telephone and cable companies is the relative success of a true "common carrier" orientation. France Telecom did not directly compete with the information service providers that used their network. Likewise, successful market makers like LAMY seek to profit mainly by bringing together buyers and sellers, rather than entering into the specific business themselves. Yet telephone companies here have aggressively pursued permission to own their own content and offer their own information services. Fears that telephone companies could use their network ownership position to favor their own services over third party suppliers (e.g., by subsidizing their services with monopoly profits, gaining marketing benefits at no cost through the use of mailings or service representatives paid for by the regulated side of the business, etc.) is the main reason that the largest online service providers in the United States maintain their own independent means of access through their own value-added networks. This further contributes to fragmentation of the market. Thus, continued efforts to define an approach that convincingly separates the provision of infrastructure from value-added service would seem to be essential to attract a critical mass of information service providers. Putting network operators into the role of gatekeeper, where they determine the specific set of information services a customer may access, will ultimately limit the formation of an electronic information services mass market.

However, prior experience does show that the network operator can and should play a proactive role in facilitating service creation and provision. This implies efforts to make it easy for service providers to join and operate profitably. At the same time, however, experiences with the "900" industry here suggests that rules to protect consumers from unscrupulous service providers are necessary. These need to be in place before users lose trust in the electronic information service marketplace, not after bad experiences destroy consumer confidance.

---

[7]This problem is now beginning to surface in the battle for control of the set top decoder market. People who obtain their "video-on-demand" service from one provider using their preferred set-top device may be unable to access another service without renting or purchasing another device. This approach makes home consumers less willing to make large investments in equipment and constrains the growth in subscribers.

CONCLUSION

Because of the complex telecommunications and information services environment in the United States, the creation of a mass market for electronic information services will not be as straightforward as many in the industry would hope. The United States cannot mimic the approach taken in France, and we clearly have benefited from the innovation spawned by competition in both network and service provision. Our analysis of the factors leading to the success of Teletel in France and audio-based and Internet services here shows the importance of universal access among both the service provider and the user populations. These requirements place premiums on policies of interconnection and interoperability. Moreover, the Internet, the Teletel system, and the "800" and "900" services industry illustrate that we need to maintain a broad definition of information service providers, with virtually any business a potential service provider.

REFERENCES

Arlen, G. (1994). Consumer online audience continues rapid growth. *Information and Interactive Services Report, 15*(8), 1.

Baer, W., & Greenberger, M. (1987). Consumer electronic publishing in the competitive environment. *Journal of Communication, 38*(1), 49–63.

Bouwman, H., & Latzer, M. (1994). Telecommunication network-based services in Europe. In C. Steinfield, J. Bauer, & L. Caby (Eds.), *Telecommunications in transition: Policies, services, and technologies in the European community* (pp. 161–203). Thousand Oaks, CA: Sage.

Bouwman, H., Christoffersen, M., & Ohlin, T. (1992). Videotex in a broader perspective: From failure to future medium? In H Bouwman & M. Christoffersen (Eds.), *Relaunching videotex* (pp. 165–176). Dordrecht, The Netherlands: Kluwer Academic.

Cash, J. I., & Konzynski, B. R. (1985, March–April). IS redraws competitive boundaries. *Harvard Business Review,* pp. 134–142.

Glascock, J., & LaRose, R. (1992, March). A content analysis of 900 numbers: Implications for industry regulation and self-regulation. *Telecommunications Policy,* pp. 147–155.

Government/industry panel examines commerical potential of the internet. (1994). *Information and Interactive Services Report, 15*(8), 4–5.

Keen, P. (1988). *Competing in time: Using telecommunications for competitive advantage.* Cambridge, MA: Ballinger Press.

Kramer, R., (1991). *Misapplying the Minitel model: The economics of information services* (Working paper series, No. 451). Columbia University, New York: Center for Tele-Information.

Mansell, R., & Jenkins, M. (1992, November). *Electronic trading networks: EDI and beyond. Conference Proceedings, IDATE 91,* Montpellier, France.

Markus, L. (1990). Toward a "critical mass" theory of interactive media. In J. Fulk & C. Steinfield (Eds.), *Organizations and communication technology* (pp. 194–218). Newbury Park, CA: Sage.

Matos, F. (1988). Information services. In *NTIA Telecom 2000.* Washington, DC: Department of Commerce, National Telecommunications and Information Administration.

Porter, M. E., & Millar, V. E. (1985, July–August). How information gives you competitive advantage. *Harvard Business Review,* pp. 149–160.

Quartermann, J. (1994). *The Internet connection: System connectivity and configuration.* Reading, MA: Addison-Wesley.

Samarajiva, R., & Mukherjee, R. (1991, April). Regulation of 976 services and dial-a-porn: Implications for the intelligent network. *Telecommunications Policy,* pp. 151–164.

Steinfield, C. (1992). U.S.: Videotex in a "hyper-evolutionary" market. In H. Bouwman & M. Christoffersen (Eds.), *Relaunching videotex* (pp. 149–164). Dordrecht, The Netherlands: Kluwer Academic.

Steinfield, C., & Caby, L. (1993). Strategic organizational applications of videotex among varying network configurations. *Telematics and Informatics, 10*(2), 119–129.

Steinfield, C., Caby, L., & Vialle, P. (1992, September). *Exploring the role of videotex in the international strategy of the firm.* Paper presented to the Telecommunications Policy Research Conference, Solomons Island, MD.

Steinfield, C., & Kramer, R. (1994). USA: Dialing for diversity. In M. Latzer & G. Thomas (Eds.), *Cash lines: The development and regulation of audiotex in Europe and the United States.* Amsterdam: Het Spinhuis.

Steinfield, C., Kraut, R., & Streeter, L. (1993, May). *Markets, hierarchies, and open data networks.* Paper presented to the International Communication Association, Washington, DC.

Streeter, L. A., Kraut, R. E., Lucas, H. C., & Caby, L. (1993). *The impact of national data networks on firm performance and market structure.* Unpublished Bellcore manuscript.

Thomas, G., Vedel, T., & Schneider, V. (1992). The United Kingdom, France, and Germany: Setting the stage. In H. Bouwman & M. Christoffersen (Eds.), *Relaunching videotex* (pp. 15–30). Dordrecht, The Netherlands: Kluwer Academic.

U.S. Dept. of Commerce. (1990). *Statistical abstracts of the United States.* Washington, D.C.: Author.

CHAPTER 14
The Long-Term Social Implications of New Information Technology

John C. Thomas
Human Computer Interaction, White Plains, New York

In this chapter, I first outline some of the major changes that are taking place in information technology and their potential social impact. These changes are classified according to whether they are essentially extrapolations of existing trends, imminent breakthroughs in new technologies, architectural changes, or meta-technological changes (i.e., changes in the way in which new technology is developed or deployed). Although any description of the future is speculative, these four cases are ordered according to increasing uncertainty. I then explore five types of reactions to the implications of new information technology. It is important to explore these reactions because the social impact of new technology is mediated by these various cognitive and emotional reactions. Briefly, the five reactions might be described as denial, anger, deal making, resignation, and transcendence and envisionment.

Aside from the limitations inherent in any prediction, this chapter suffers a further limitation in dealing specifically with social implications. Social phenomena are necessarily complex and attributions difficult to ascertain even in discussions of what has already happened. Thus, depending on one's agenda and experiences, a whole variety of different "reasons" can be advanced for the rising divorce rate. People have blamed it on everything from feminism to the automobile to a communist plot mediated by rock music. We cannot "redo" the past under a number of controlled experimental conditions and so each attributional hypothesis remains just that: a hypothesis. In dealing with the future, we can only make reasonable guesses about how technology will evolve and how that will interact with social trends. Another limitation of this chapter is that it primarily looks at how new technologies will impact "mainline" culture in the United States.

Despite these limitations, the exercise of looking at possible social implications of new technologies is felt to be better than the alternative of blindly building new technologies heedless of their human impacts. The reader may agree or disagree with various portrayals in this chapter; may find some of the implications frightening or reassuring. In any case, it is hoped that everyone who participates in building new technology will spend some time thinking about how that technology might affect all of us.

THE NEW INFORMATION TECHNOLOGIES

New information technologies may be categorized as trends, breakthroughs, architectural changes, and metatechnological changes.

Trends

Most readers are familiar with what can be characterized as trends: increases in bandwidth, decreases in prices for memory and compute power, smaller power requirements for computers, and so on. Although the exact points on some of these curves is difficult to predict far into the future, barring some ecological, cosmic, or political disaster, we can be fairly safe in assuming that the technology of 2020, for instance, will provide computers that are much more widely deployed and available. “On-ramps” to the “information highway” will be everywhere. Power requirements, size changes, and cost changes in both CPU and storage will allow the power of today’s workstation to be found in an index card, a wrist watch, a pen, or a magic slate. Decreases in the cost of bandwidth will also mean that these powerful devices will basically serve both as local intelligence sources and as gateways into much larger and more powerful computers and data bases as well as interpersonal multimedia communication networks. We are already seeing an explosion in the use of phones in cars and on planes. Such trends will continue and accelerate so that we will literally be connected to a web of information almost anytime we wish (and maybe when we don’t wish!).

Breakthroughs

We expect to see substantial advances in identification technologies (the ability of a computer to recognize a person from handwriting, speech, and other behavioral patterns or physical characteristics) as more is learned about recognition, neural networks, and other machine learning techniques (see, e.g., Alspector, Goodman, & Brown, 1993). Larger databases of examples will also help improve these technologies. In addition, we may well see breakthroughs within recognition technologies such as the ability to deal more effectively with nested contexts. On the positive side, carrying a whole host of various credit cards along with cash may well be unnecessary. All the responses of the computer web could also be personalized to you and your preferences. On the negative side, it may well be the case that you can be “tracked” or “tracked down” throughout the planet.

Similar to identification technologies will be transcription technologies; that is, translating amongst various media of communication. Speech recognition allows the computer to translate spoken speech to data; speech synthesis to produce words from ASCII; handwriting recognition allows transcription from handwriting to an internal representation and so on. These technologies will allow easier to use interfaces in many applications. Beyond that however, what they collectively imply is that almost any human communication, no matter how informal, may be transcribed automatically into data. In combination with identification technology, it basically means that what is being written, said, gestured, and so on, and who is saying it may be recorded.

Such technologies will also allow us to see various kinds of trends and patterns that are beyond the scope of one individual to note. For example, one could literally "follow" the course of a rumor or a new idea around an organization. One could see long range spatiotemporal changes in language behavior. One could "map out" a productivity curve for various kinds of activities as a function of time, weather, or other factors. Will the information thus obtained be filtered in some way to protect privacy? Will it be available to everyone in society so that empowered decentralized decision making will be facilitated? Or, will such information be a centralized resource that is ostensibly used to improve "efficiency" but actually used to consolidate power bases?

We may also expect that many "transcriptional" tasks will be performed by machines and the human tasks that remain in several decades will deal almost exclusively with the exceptions and the tough cases. Because people will be dealing more and more with exceptions, education should begin now to stress problem solving rather than routine. People in the service sector may find these changes in job structure interesting and fun, but much of that outcome will also depend on how new technology is introduced. To the extent that workers have some control over their lives and are able to communicate in the way that they wish, such jobs will be more rewarding rather than simply more exploitative (cf. Edwards, 1979; Garson, 1988).

Although having routine transcription done by machine will be more efficient, it will also combine with identification technology and cheap storage to mean than nearly everything communicated can be saved and categorized. Not only might "big brother" be watching; lots of "little brothers" might be watching too! Even today companies are beginning to keep track of buying habits through the use of bar-coding combined with "discount cards" that identify who you are. Arguably, this could result in better service for the customer, but it is interesting that this use of discount cards has not been widely publicized. If identification and recognition are universally applied, companies could "listen in" on every sales discussion. They could find out not only what you bought but why you bought. This could be used to make products that better suited the marketplace but also used to help train more effective sales people (see Katz, 1990, for an excellent discussion of privacy and communication). Davis and Botkin (1994) gave positive examples of how this technology may help customers (e.g., "if a guest at the Boston Ritz-Carlton requests six allergenic pillows, the hotel's knowledge-based system will make sure she finds them when she checks into any Ritz in the world").

We are now seeing the beginnings of another set of concepts known as virtual reality (VR), augmented reality, cyberspace, and virtual environments. As computing power gets more powerful, the illusional capabilities of these technologies will become greater. VR may have applications in entertainment, education, teleoperations, and surgery. Perhaps the most profound long-term applications, however, of such technologies may be as an "empathy machine" (Stuart & Thomas, 1991). Perhaps by 2020, it will be possible to very quickly give one person a "feel" for what it is like to be someone else or at least someone in a different role or context. To take a simple example, suppose that an adult could "see" what it is like for a child to be violently handled by a huge adult or how the world looks and feels to their elderly parent. What effect would such "empathy enhancement" have?

At best, we could imagine that it could serve an important purpose today and one of growing importance tomorrow. Such an empathy enhancement machine could be more important because of the growing diversity of experience due to greater job specialization and the individualization of communications. Unlike other media that can be used for empathy (e.g., film, novels), VR has the advantage of potentially being interactive; the user actually sees and feels consequences of their own behavior. This may make it the most effective medium for empathy yet devised.

Another breakthrough technology already on the horizon is that of intelligent agents (e.g., see Rosenschein & Zlotkin, 1994). People will have at their disposal, computer programs that operate in some limited domain much the way that they would operate. For example, many people already use intelligent "filters" for their e-mail (see Fischer, Lemke, Mastaglio, & Morch, 1990, for further discussion). A more recent development along similar lines is the use of "robots" or "bots" to respond to certain keywords in chat rooms or e-mail discussion groups with standard comments. One possible social implication is that people may experience less and less trust; they may respond only on the basis of the "face" that is presented with no real attention to an underlying personality (because there might not be one!). Stone (1991) told of a case of a male doctor who "impersonated" a female via e-mail and held extremely intimate (and supposedly helpful) conversations with females. When it was later discovered that "she" was "really" a "he," many of the female patients felt very betrayed.

Taking artificial intelligence (AI) in a slightly different direction, by combining intelligent agents with identification technologies, one can begin to operate "personalized" software. Already, there are intelligent tutoring systems (ITS) that induce something about the particular student's state of knowledge and tailor lessons accordingly (Gray & Atwood, 1991). We may soon expect that computer programs more generally, including computerized communication channels, will "recognize" who we are and give us different versions, different defaults, different parameters, and different styles of interaction. This may have obvious advantages in productivity and personal interest, but it may also have the effect of having various individuals using a system having less in common. This could adversely affect productivity because people would not be able to share information on how to use systems as effectively.

Such a development may have even more profound effects on social interactions as it becomes less useful to seek another individual's help since that other person's computing and communication world would be so different. There could be an isolating impact of extreme personalization. There has already been a significant change in the commonality of television and movie communications in the last few decades. In the 1950s, 1960s, and 1970s, most people watched network television or saw one of a few movies that were showing at the cinema. Now, people have access to one of many channels of cable television and can rent any one of hundreds of movies at their video store. Soon, such choices will be available via the network. By 2020, we could also expect that even these communications could be further personalized. There may well be a chip in my set-top box that will allow me to see a movie with the type of endings I like and possibly even with my favorite actor's (or even my own) image visually "dubbed in." Again, the net effect may be that various individuals in a society come to experience very different things. This in turn could negatively impact interpersonal communication and a sense of unity.

Architecture

In addition to the growth trends in the computing and communication (C/C) industry and the introduction of new technologies such as those just mentioned, the industry will also see a diversification of architectures that will further multiply the effective power of C/C systems. In the early days of computing, for example, machines were built according to a "von Neuman" architecture in which items were stored in and retrieved from a passive memory and acted upon sequentially in a central processing unit (CPU). Although such an architecture makes it relatively straightforward (although tedious and time-consuming) to program and follow what is happening step by step, what is strange about it is that very little of the computer memory is active at any point in time. If one were to draw an analogy to a corporation, it would be as though only one worker in a company at a time were allowed to do any work. The executive would command a particular worker to do a particular thing while every other worker stood still. Then the next worker would do their job and so on. What is clearly of paramount value in such a scheme is that everything is understood and proceeds according to an overall plan; everything is under control. From a hardware perspective, all the memory elements were the same and the central processor was the same for all tasks. What made the computer do different things was that different sequences of instructions were programmed into it.

Today, we are seeing the reintroduction of specialized hardware. There are specific kinds of memory for displays, for instance. There are special kinds of CPUs for floating point arithmetic. Parallel computer architectures are becoming more widespread. In the future, we can expect to see even more diverse architectures. A computer will be a collection (community?) of machines, each of which is designed to do certain kinds of task extremely well. We will also see hierarchically and heterarchically organized sets of specialized systems of control (more like animal neuronal mechanisms; cf. Brooks, 1990). In some cases, the memory itself will serve as an active element. For example, in today's architecture, in order to provide updated displays for a virtual reality system, as the user moves their head to the left, points in the visual field stream by in a very predictable fashion. In order to calculate the new display, every point is assigned coordinates, new coordinates are calculated by formula and these coordinates are used to paint a new display.

In the future, we could expect to see specialized hardware for rotation, displacement and streaming in which active memory elements automatically "calculated" next points simultaneously for the whole visual field (cf. Furnas, 1991). The impact of architectural changes in computing may largely be in making computers more efficient and thereby enhancing their power over and above what one would expect even in terms of the historical trends of smaller size, lower cost, and higher capacity mentioned earlier. But there may also be a more subtle impact on the sociology of computing. In the past, most computers were built along the same general lines that favored a particular kind of linear, mathematical thinking. To the extent that general purpose computers were used in art, music, and other endeavors, problems had to be formulated in terms of these linear processes. If architectures of the future, however, are specifically suited to a variety of types of information and processes, we can imagine that the thinking that goes into building an application program, say, for music will be much closer to the ways musicians themselves think about their field apart from the need to "computerize."

In turn, this may imply that, the fundamental design of computer software and hardware may be much more the province of people who are professionals in a given application domain, rather than computer professionals. In turn, the computer designs that result will reflect a much broader cross-section of the population than is now the case. This will hopefully have a salutary effect on the entire computer industry.

Other interesting effects may come about in communication architectures. Little more than a decade ago, the public phone network in the United States was largely dominated by one company, AT&T, which was responsible for local phone connection, long distance calls, designing networks, building switches, building handsets, and in general, everything having to do with the provision of telephone service. Whatever the merits or demerits of that system, the situation today is one of fierce competition in every arena. The architectural challenges are much more complex because we now have multiple interlocking architectures wanting to provide multiple services that deal not only with the transmission of voice, but also with data, video, and multimedia that can be not only transmitted but also stored and transformed. These trends feed into the personalization and ubiquity trends mentioned previously, but deserve special attention if the plethora of possible services provided by a myriad of providers is going to offer a coherent suite of functionality from the end-user's perspective. Otherwise, the "information highway" will be too confusing to enter.

Another aspect of new C/C technologies is the style that such artifacts impose. These styles, in turn, may well impact society apart from the content of messages or the technological possibilities. For example, it has been posited (Landman, 1986) that the speed of computing has contributed to the cult of celerity. Because people get such fast responses from computers and computers, in turn, make up such a large proportion of their interactions, people come to expect all other human activities to proceed on such a time scale. People, to put it succinctly, are impatient. They want to cut to the chase, get to the bottom line.

Similar arguments were made earlier that computers, because of their dichotomous (0 or 1) natures push people toward similar dichotomous thinking. Sometimes this is useful. At other times, the search for which of two answers "is" right blocks the ability to even conceive of a more creative "out of the box" thinking process. "If the computer can't take the information in that form, it doesn't exist." One person, for instance, joining a large corporation from another culture had to "manufacture" a name because the computer system "wouldn't allow" someone to join the company with only one name. Handwriting and speech recognition systems typically work relatively well on a large proportion (about 80%–90%) of people (called in the recognition literature, "the sheep") but work relatively poorly on a small segment of the population (about 10%–20%, "the goats"). Life becomes less convenient for such people. What are the implications? The choice of which technology and how people communicate within it impacts our perception of others (Thomas, 1983). People who chose "formal" as opposed to "informal" means of communication, for example, were seen more negatively on a number of personality dimensions.

Metatechnology

Changes of an even less directly "technological" sort that may profoundly impact the society include how we conduct business. Humans have learned to operate fairly well in small groups or teams. We are less proficient at how to make large corporations, governments, and universities operate anywhere nearly so effectively as the sum of the parts might imply. This is not surprising because the interactions, communications, and complexity of such systems are far greater and as a species we have been relating in small groups for millions of years but in large enterprises only for a couple millennia. Perhaps the first models for such large enterprises were building pyramids, harvesting, and waging war.

The models for such activities worked when a plan could be centrally organized for a stable environment and carried out by more or less interchangeable masses of people. In the current world of rapid change and diverse such systems do not work very well (Peters, 1993). Large corporations are experimenting with a variety of procedural, technological, and even attitudinal and spiritual approaches to making their organizations work (e.g., Imparato & Harari, 1993). We see innovations being tried in universities and governments as well. To the extent that our major institutions begin more fully utilizing their personnel, consequences could far surpass anything caused by technology. For example, professionals in the field of human computer interaction have noted for some time (e.g., Chapanis, Garner, & Morgan, 1949; Fitts & Jones, 1947) that in order to design an effective new system that has both human and machine elements in it, one must consider both the human and the machine elements. In order to get a truly productive system, it is necessary to iteratively design and implement with the input of real users.

Landauer (1995) looked at the correlation between how much money various industries have spent on computerizing over the last few decades and their rise in productivity. The correlation of these expenditures with increases in productivity is basically zero. The vast majority of these systems (misnamed computer systems because they are really HUMAN–computer systems) have been built and deployed from a perspective of thinking about the computer alone. In those few cases where iterative design was used with the input of real users, the annual productivity rate improvement averaged 30% rather than 1%.

In addition, more attention to such "human" factors could help prevent tragic accidents (Casey, 1993). Iterative design of systems with input from users is just one example of how corporations might do things better. Others might include using multimedia to provide a corporate memory of what has worked and what has not worked over time. Perhaps having "automatic" tracking of activities could allow, in conjunction with more sophisticated computing, decision making based on a more complete cost–benefit analysis rather than a cost basis. Today, costs are much easier to measure because they are more localized in time, space, and organizational partition than are benefits.

To summarize briefly, there are a host of ways that new technologies will impact our society. First, the log-linear trends in the cost of computing and communication will mean that people will have access to huge amounts of information from virtually anywhere at a small charge. Second, there are new breakthrough technologies such as identification technologies that will allow people to be "tracked" as to location and activity. Third, there are changes in architecture that may magnify the impacts of the trends just mentioned. Fourth, there may be profound social impacts due to the style of our new technologies. Finally, during the next few decades, large organizations may learn to become much more effective in how they operate including especially the use of new technologies. If this happens, another multiplier effect will be applied.

REACTIONS TO NEW TECHNOLOGY

Although we have speculated about many technological changes, in order to understand better what impacts these changes might have, we need to examine the kinds of reactions that people exhibit to change. Change does not impact adaptive intelligent systems like individual humans, teams, corporations, or societies in an unmediated fashion. Much of the impact depends on how people "take" the new technology. I now examine various kinds of reactions in turn.

Denial

In denial, individuals basically say: "These technologies may be all very interesting but they won't really change things." A slight variant of this, of course, is to allow that these new technologies may change other people's worlds, but not mine. A doctor in this phase might well admit that television will change the entertainment industry, but not medicine. A teacher might well think that TV is great for entertainment, medicine, and even religion, but that the teacher in the classroom is still the most effective way to teach and will never be replaced. A minister, on the other hand, may well feel that television has many implications for education, entertainment, and medicine but obviously has nothing to do with religion. In hindsight, of course, we see that all of these reactions are wrong. Television has affected nearly every aspect of life, especially in America. In fact, it has even been argued, that television has affected books. According to Postman (1986), Americans have gotten so used to soundbites that even books must be formatted into little two or three page chapters in order to be readable by an increasingly impatient public. By contrast, less than 150 years ago, ordinary citizens listened all day to the Lincoln/Douglas debates.

Another form of denial is the belief that these technologies may eventually change things but that no action is necessary. With all the recent press about the Internet and the information highway and all the press in the last few years about VR, it may be thought that few Americans are in the denial about new information technology. There are still many politicians, however, who believe the United States can live in an isolationist economy, and that we can compete in a global economy without using computers or automation to make our workforce more productive. In fact, at the White House itself, in January 1992 the President of the United States did not have a touch-tone phone and phone calls were switched by human operators in the basement using plugs of 1930s vintage.

What are the social implications of having one large segment of the population embracing new information technologies and attempting to bend them to human purposes while another large segment denies that such technologies have much effect, or alternatively, that they may have an effect but such effects will take place independently of any actions on their part? One potential is a bifurcation of society into the informational "haves" and "have nots." A corollary of this might be increasing dissension of voting and governmental gridlock between those caught in defending against new technology and those embracing it.

Anger

Here, new information technologies are seen as impacting society, but only for the worse. Postman (1986) made a fairly reasonable case that television is a major cause of much that is negative in U.S. society. More recently, he turned his attention to the computer. Postman (1994) claimed:

> In the case of computer technology, there can be no disputing that the computer has increased the power of large-scale organizations like military establishments or airline companies or banks or tax collecting agencies. . . . But to what extent has the computer technology been an advantage to the masses of people? . . . These people have had their private matters made more accessible to powerful institutions. They are more easily tracked and

controlled; they are subjected to more examinations, and are increasingly mystified by the decisions made about them.

It is probably true that the combination of computers and communications make it more difficult (but not impossible) for people to cheat on their taxes, avoid paying their parking tickets, and get hired for child care in Texas despite some molestation convictions in Tennessee. It is also the case, that through the misapplication of computers, people are denied credit (e.g., when they really should not be). If, for some reason, a computer system is confused about who you really are, even if you are totally innocent, you may be the victim of bad data in the computer systems used by universities, governments, and corporations. It seems likely that the solution is not to give up all the computers in the world, but to use what we know about systems design to make sure that such systems are useful, usable, and humane; that people are taught to believe that just because a computer says it does not necessarily make it true.

Computers do not just provide benefits for large organizations. They may also provide greater productivity and economic benefit as well as convenience to individuals. The IRS may use computers to help track down tax fraud, but many individuals use computers to help with tax planning and preparation. Airlines may use computers to help make reservations in such a way as to maximize profits, but without a computerized reservation system, making flight reservations would be a confusing, time-consuming process for the individual, and the ticket prices would be higher, to reflect lost productivity. It is true that over the last decade in the United States, the rich have gotten richer and the poor have gotten relatively poorer. But surely this has more to do with an explicit political and economic policy with precisely the aim of "Trickle-down economics" than with the fact that computers are more powerful, cheaper, and more ubiquitous. Computers are also widespread in Scandinavia, for instance, where a different political and economic policy has achieved different results.

There are trade-offs that occur with any change. There are many instances of problems that exist today (e.g., automobile accidents, software-caused accidents) that could not exist without the underlying technologies. But to rail against new technology on the basis of comparing the negative accompaniments of the new technology without a look at the benefits and by comparing it against an idealized and romanticized past would seem to accomplish little in terms of the very real and very important task of directing new information technology to maximize human benefit and minimize human pain. Perhaps one must pass through a stage of unreasoning anger in dealing with changes in technology, as in the news of one's impending death, before one can progress to a more constructive view of things.

Of course, not every technological innovation has more potential for good than bad. In some cases (e.g., germ warfare) it is difficult to see how to turn the technology to a positive purpose. Even more ubiquitously, new technology, as stated earlier is often designed and deployed in a context that does not adequately appreciate user's tasks, contexts, and physical and psychological needs. In these cases, the "new technology" does have a negative impact. Anger is certainly an understandable response then; we might hope though that the anger is turned to the constructive and mutually beneficial task of ensuring that new technology is redesigned with the human beings in mind. After all, technology should be a means to increase human happiness, not an end in itself!

Making Deals

People may accept that new information technology will come and that it will have effects, but try to "make a deal" with reality to limit the scope of that change. For example, a typical deal is that people may well realize and accept that computing technology may make things more efficient. In other words, it is all right to use technology to do what we do now but do it faster or for less cost. In terms of our major institutions, the idea is that schools, banks, businesses, prisons, churches, and so on will continue to exist in the way that they now exist and they will continue to do what they now do. But new information technologies will allow churches to become more efficient at gaining converts as well as doing the business of the church. Telephone operators will give information more quickly. Prisons can use surveillance technology to more effectively surveill. Banks can shift money around more efficiently and make even more money. This is probably the reaction that most people in the United States have with respect to the new information technologies.

Acceptance

People may also accept new technology; they realize that new information technologies are going to change what basic institutions are. This will cause the very character of human life and humanity to change. This is not to say that we will have no connection to our past. Our preliterate ancestors probably enjoyed a good meal, touch, and a warm fire. This will not disappear with the advent of new information technologies. But as today, some aspects of our social life hearken back to a small tribe; there are other aspects of our social life that are very different because of the technologies that are already widespread in our society. Suppose that in a few decades, the new technologies discussed in this chapter are as widespread as the TV, telephone, and automobile are today. Our social lives may change as much as a consequence in the next 25 years as they have between 20,000 years ago and today.

For example, the very institutions that we take for granted as being a part of our society: Churches, families, schools, prisons, governments, and companies may not even exist as such. We see already the appearance of such terms as *infotainment* and *infomercials.* To take the example of colleges, we see a boundary blurring with corporations. Many courses are taught by companies. Many teachers at college are adjunct faculty. Courses are dictated as much or more by the needs of the job market than by any academic abstract notion of what constitutes an education. Universities have marketing departments and worry a lot about profit and loss. Many colleges and universities as a part of their programs have internships in real-world settings. Academics typically no longer say, "that is not true" but rather, "I cannot buy that." At the same time, education is attempting rather heavily in many areas to be more "entertaining." Reading is being replaced by video. Laboratory experiments are being replaced by computer simulations in some cases. Some things are lost with these changes; some things are gained. But the changes, we must recognize, are profound. It is not simply that colleges are using the new information technologies to become more efficient. What a college is and is about is changing.

Transcendence and Envisionment

In this section I take a different view of the social impact of new technology; rather, I reframe the issue away from the passivity implied in the word, *impact* and take a more proactive view. Let us begin by examining some of the major unsolved problems of the world and then consider how new technologies may help solve these problems. The major problems that people face might be categorized as "Relation to self, relation to the physical world, and relations with others." At first blush, only the last of these is primarily, indeed, definitionally social in nature. A moment's thought however, shows that all of these types of problems impact our social world. Criminal behavior, depression, alcoholism, drunk driving, racism, drug addiction all have profound social impacts as well as partly social causes although they are certainly manifest as a problem in the individual's relationship with him or herself. Similarly, world hunger, pollution, and the destruction of the rain forests are basically problems of the relation of humans to the physical world but ones that have profound social consequences. Despite these interdependencies, it is useful to examine these major areas in turn to examine how new technologies might address human problems.

Problems in the Relationship to Self. Smoking kills. Yet, people continue to smoke. This seems curious but it is not unlike many other behaviors that people engage in that seem nonoptimal if not outright self-destructive. The fundamental problem in ordering our behavior is this: Rationally, we can see the impacts of our behavior across time and space and persons. But, in terms of our ability to change our behavior (or habits, or attitudes), we are primarily impacted by what happens very close in time and space. For example, if we stop smoking, we feel worse immediately. If we have the cigarette, we feel better. This short-term hedonism overrides the longer term gains of not smoking.

For most animals, such situations provide no paradox. Animals learn and react according to short-range hedonism, but they do not "know" any better. People's learning and emotional reactions are often in fact governed by principles very similar to those that govern other animals. However, people also have an ability to imagine further reaches in time and space. The imagining of some future gain however, is not typically nearly so strong as an actual near-term stimulus in controlling behavior. If we posit that our main difficulty in controlling our own behavior is in using long-range (over time, space, and others) rather than short range hedonism to control our behavior, then what can technology do about this?

First, we must recognize that technologies to deal with this problem have been evolving for thousands of years. Commandments written on stone tablets, "to-do" lists, pictures of thin people on the refrigerator door, Caesar's "pep rally" speeches to his troops before battle, TV images of what is happening elsewhere in the world, pictures of the gray lungs of smokers or the diseased livers of alcoholics—all these things are attempts to use communication technologies to help influence our behavior across greater reaches of time and space.

Control techniques include using negative stimuli, positively reinforcing incompatible behavior, and reducing the negative impact of avoiding the behavior to be changed. The basic problems with existing techniques that rely on providing negative stimuli associated with behaviors to be extinguished are: lack of timeliness, lack of perceived relevance, and lack of vividness. The "lack of timeliness" is this: To be effective, a negative stimulus must be available at the time of behavior. Showing a picture of a gray lung on Tuesday will have virtually no impact on my smoking on Wednesday. Visual stimuli, in particular, can be highly ineffective because we may consciously shift our gaze and visual attention. Thus, I can easily pick up a pack of cigarettes and simply choose not to read or even look at the surgeon general's warning.

Lack of perceived relevance comes about because of the fact that not everyone who smokes dies of lung cancer. People who smoke may well convince themselves that, for example, because they also take vitamin C or eat broccoli, they will not get cancer.

Now, the question is whether new technologies can be used to effectively control negative human behaviors. The answer is: "yes, but. . . ." The most obvious way to prevent alcoholism, smoking, and perhaps violent behavior is simply to put each child in a controlled medical setting at an early age, have the child associate the smell and taste and feel of these states with extreme nausea (not unlike what was done in Stanley Kubrik's *A Clockwork Orange*). We tend to think of this as far less civilized, however, than giving everyone free choice and at some later point in time allowing some proportion of people to die of lung cancer, chirrosis, or capital punishment.

Another, more costly alternative is to provide universal monitoring with the new technologies and associate that with credible, costly threats. Thus, you can smoke, but if the sensors catch it, you are not fined a few bucks, but lasers slice your hands off. Such a solution does not really require new technology however. If police officers were given the duty to simply shoot on sight anyone smoking or drinking, these specific problems would cease to exist very quickly.

The point of this thought experiment is certainly not to suggest that we implement such Draconian plans, but to make it quite clear that society already has the means to eliminate many of the problems that people have in relationship with themselves via tightly controlled negative stimuli either in the form of associative learning or through threats. We do not choose to implement such plans, not because of lack of technology but because we feel the social cost of such measures outweighs the benefits.

Another path to consider is to put the delivery of negative stimuli under the control of the individual but to use new technology to make these stimuli more vivid, more timely, and more relevant. To continue the smoking example, the smoker who wants to quit could down-load a tear-jerker movie about someone dying of lung cancer every time he or she was tempted to smoke. Further, the actor's face and voice could be substituted via a personalized program with the person's own face and voice allowing them to "identify" with the character in the drama even more closely. Via the ubiquity of connection, such multimedia personalized stimulation might be made available virtually anywhere. With gesture recognition and smoke detectors, we could also imagine that this *infotainment?* could be delivered just as person began smoking.

In addition, the program could be "time-locked." In other words, once the person decided to quit, the programming could be delivered to them without their being able to override it for 6 months. Such presentations could be made even more effective by requiring them to be interactive. For instance, the individual would not be able simply to look away, not listen, or otherwise ignore the messages because active responses would be required of them. Only when the "correct answer" was given could they avoid punishment or gain reward of some type.

Even though the individual began this program on a voluntary basis, many people may feel as though this scenario is far too coercive. This is not primarily a technological problem but springs from a much deeper problem. We like to feel as though people have free will and that a person should be able, for instance, to quit smoking on a free, rational basis. In fact, it is difficult to do this, so we long for better techniques and technologies to help. But when we find one, precisely to the extent that it is universally effective, that is taken as evidence that the technique is coercive—taking away the person's freedom of choice. Analogous scenarios and arguments can be made for overeating, underexercising, antisocial behavior, drug abuse, and so on.

Perhaps we could extend the concept to the worker who feels he or she is not being productive enough. This last example leads, of course, to one of the real difficulties with coercive technologies, even when self-imposed. How does one prevent people with power (e.g., employers, insurance companies, governments, admissions boards to colleges) from telling people they "have to" undergo treatment before they will be able to get a raise, get insurance, get their license renewed, or be readmitted to college? Any program that appears to be voluntary can be made involuntary through the application of external power. This is one of the reasons for our mistrust of truly effective techniques of behavioral change. The conclusion then, is that the direct application of new technologies to try to make behavioral control more effective will not work, even if it appears on the surface that such control is in the hands of the individual. A prior challenge that must be met is greater interpersonal trust and trustworthiness.

Greater communication bandwidth, better computer-supported cooperative work (CSCW) tools, and various ways of using new technology to enhance empathy and understanding may eventually reduce human hostility and suspicion to the point where people may be willing to use highly effective means of self-control. Until such time as these technologies do effect a profound change (or until such a change comes about through other means), the prospects for using such technologies are dim: Better self-control through technology seems too dangerous despite the huge potential benefits.

Problems in the Relationship to the Physical World. Humankind's relationship to the physical world has, in turn, profound shaping influence on our social relations. Indeed, to take an extreme example, imagine how different social relations would be if there were enough material wealth for everyone to be completely satisfied. The impact that new technologies will have on our relationship to the physical world is difficult to predict. On the one hand, population growth is continuing to outstrip our ability as a planet to feed our population. We are using up many of our nonrenewable resources at an increasing rate. Although there have been some recent advances in the ecological arena, it is still quite possible that we have already embarked on a path that will make the planet literally uninhabitable by humans within a century for any one of a number of reasons. On the other hand, advances continue in virtually every field of science including especially agriculture, geology, genetics, and ecology as well as computing and communications.

Faster, smaller, lighter cheaper computers connected via high bandwidth in a worldwide web may well reduce the necessity of fuel consumption due to travel. As we use computers to work more efficiently, we may well reduce further the utilization of nonrenewable resources. As mentioned earlier, in order to realize the potential benefits of such new technologies, it will be necessary for human considerations to be part of the design process to a much greater extent than is typically the case today.

If this happens, what we take as normal "human nature" itself may radically change. People may become quite a bit more generous and trusting, for instance. This in turn, could add further to planetary well-being. Imagine the cumulative social impact, for instance, if every time a customer called to find out information from a company, it was a successful and pleasant experience both for the customer and every worker who was involved. Imagine the cumulative impact of eliminating most of the traffic delays and inconveniences that many commuters now face through tele-commuting and better dynamic routing. Imagine that people who were customers at stores, employees at corporations, patients of doctors felt that they were empowered and contributing partners. Such changes would provide social benefits over and above the productivity gains; less hostility would pervade social interactions. People would begin to have a positive expectation about such interactions; feel comfortable to share more information; help each other out.

Perhaps the new C/C technologies could help establish a wide sense of community. Even subtle changes in computer-human interaction can alter a user's experience significantly (Walker, Sproul, & Subramini, 1994). Because the science of human-computer interaction has been applied in such an uneven fashion, it is not surprising that attempts at automation have had mixed results so far in subjective effects (cf. Kraut, Dumais, & Koch, 1989). Even subtle effects make a difference, thus we might expect that systems designed according to "best practices" might have profound subjective effects (as well as increasing productivity).

Problems in the Relationships Among People. Natural language communication between human beings is often problematic. There are many useful mechanisms in language (Thomas, 1978) but there are also many cases in which human–human communication shows breakdowns. In Thomas (1980), several root causes of these breakdowns were outlined. One root cause is that people focus on differences. Another is that natural language evolved in situations wherein people shared a common framework and background but is now used in situations wherein people do not share common frameworks and backgrounds. A third problem is that the repair mechanisms for communications breakdowns are often aimed at finding the person to blame. Because communication (and communication breakdown) is an interaction effect, it is impossible to find out who is to blame. Hence, naive attempts to fix communication breakdowns by finding out who is at fault necessarily escalate negatively. Let us examine these three difficulties in turn.

An example may help clarify the first difficulty: focusing on differences. Our perception is geared toward focusing on differences and in most circumstances this is the right way to allocate limited resources. As hunters, gatherers, proofreaders, critics, inspectors, recruiters, and most of our other roles, we need to focus on differences among objects. However, the application of this generally useful heuristic to the special case of our perception of other people results in rather ridiculously one-sided views of our relationships to each other (and to nature, for that matter). Human history has shown prejudice based on any number of differences, most of them rather trivial compared with the similarities.

One obvious way that technology can be used to help us focus on similarities is to use it to "filter" out differences. In online chat groups, for instance, you cannot see a person's skin color, sex, or age; you cannot hear their accent. Nonetheless, among the most common questions asked of people entering chat rooms is their age, sex, and location.

A more active way of having technology help with this root cause of prejudice is to have a representation of people that separately lists the similarities and differences. Because information is still costly and we have crude ways of scanning through it, most information about a person (e.g., resumes, online profiles, personnel files) focus on differences. If we had a very effective and efficient way of storing, retrieving and organizing information, we could imagine a two-step sort of resume, profile, or personnel file that first reminded us about what was true about this individual as well as everyone else. Then, we could look at the unique information.

A special case of this focus on differences has to do with our relationship to our goal structures. We may be working with others in a team, a company, a government, for instance, and share a whole host of top-level goals. If however, we are solving particular problems, we may well focus so heavily on the goal at hand and on the differences of goals at that lower level, that we completely ignore all the higher level goals that are shared. This may even take place to such an extent that the team is destroyed. Technology might be able to help by helping us keep simultaneously in mind some auditory, tactile, or visual information that shows the entire goal structure in priority order as a counterbalance to our natural tendency to overfocus on the differences.

Let us now turn to the problems brought about by having people work together across many different frameworks and backgrounds. Spoken (and even written) language, being linear, are not especially well-suited to helping people understand the often implicit background within which someone else is operating (cf. Thomas, 1978). Film, by contrast, seems a better medium for showing a whole context. Potentially, easily accessible multimedia in conjunction with intelligent agents, may be capable of helping people understand each other's contexts. Broadband communication may also help people stay aware of each other's contexts more effectively. By being able to see the visual environment of another as well as hear what they are saying, we may well be able to stay aware of important background factors such as the weather, other activities, the person's mood, and so on.

Dealing with breakdowns might also be aided by technology. For example, an "active" computer system on the communication channel may help coach people to analyze the system of communication and help them design a better one rather than focusing on "blame." It has been suggested that focusing people on the process of designing a system, attention is diverted from the confrontation to a common goal (Kellogg & Thomas, 1993).

SUMMARY

Obviously, predicting exactly how the future will turn out is impossible, even in broad outline. Nevertheless, by examining the issues, it is possible to clarify the probable impact of current choices. This chapter has examined some of the likely technological trends and their social counterparts. We have also noted that the reactions to new technologies are not universal but depend on how individuals and groups view the technologies. We have looked at various proposed ways people deal with the impact of new technology: denial, anger, deals, acceptance, and transcendence. It is my hope that individuals, governments, and corporations will largely take this last attitude. New technology has always offered danger as well as opportunity for a better life. This will continue. There are actions that can be taken today, for example, designing systems with people in mind that can have a profound impact on the chances that new technological systems will result in happier, more productive people. In the best case, we can imagine that, to the extent that people are basically more good than evil, that greater communication will eventually result in more good in the world; that greater bandwidth and access to information more effectively presented will result in greater worldwide as well as local cooperation.

A story is told of someone who wanted to see heaven and hell before death. A god granted his wish. He was given a glimpse of hell. At first, he was surprised because what he saw was a room with a rich feast laid out on a huge circular table. Then he saw that it was indeed hell because each would-be feaster had a fork that was too long to reach his mouth. The people were starving amidst plenty and in obvious agony. Next he was shown heaven. It too was a room with a rich feast laid out on a huge circular table. Again, people had forks too long to reach their mouths. But here everyone was jolly and fat.

"I don't understand. Heaven looks the same as hell."

"Yes, replied the guide, but you see, in Heaven, people have learned to feed each other."

ACKNOWLEDGMENTS

The author thanks Wendy Kellogg, Beth Adelson, and Ruby Roy Dholakia for comments on earlier drafts of this chapter.

REFERENCES

Alspector, J., Goodman, R., & Brown, T. X. (1993). *Applications of neural networks to telecommunications.* Hillsdale, NJ: Lawrence Erlbaum Associates.

Brooks, R. A. (1990). A robust layered control system for a mobile robot. In P. H. Winston & S. A. Shellard (Eds.), *Artificial intelligence at MIT: Expanding frontiers* (pp. 2–27). Cambridge, MA: MIT Press.

Casey, S. (1993). *Set phasers on stun: And other true tales of design, technology, and human error.* Santa Barbara, CA: Aegean.

Chapanis, A. Garner, W. R., & Morgan, C. T. (1949). *Applied experimental psychology: Human factors in engineering design.* New York: Wiley.

Davis, S., & Botkin, J. (1994, September–October). The coming of knowledge-based business. *Harvard Business Review,* pp. 165–170.

Edwards, R. (1979). *Contested terrain.* New York: Basic Books.

Fischer, G., Lemke, A. C., Mastaglio, T., & Morch, A. I. (1990). Using critics to empower users. In *Proceedings of CHI 90.* Reading, MA: Addison-Wesley.

Fitts, P. M., & Jones, R. W. (1947). *Psychological aspects of instrument display. I: Analysis of 270 "pilot-error" experiences in reading and interpreting aircraft instruments* (Rep. No. TSEAA- 694-12a). Dayton, OH: Air Materiel Command, Aero Medical Laboratory.

Furnas, G. W. (1991). New graphical reasoning models for understanding graphical interfaces. In *Proceedings of CHI 91* (pp. 71–78). Reading MA.: Addison-Wesley.

Garson, B. (1988). *The electronic sweatshop.* New York: Penguin.

Gray, W. D., & Atwood, M. E. (1991). Transfer, adaptation, and use of intelligent tutoring technology: The case of Grace. In M. Farr & J. Psotka (Eds.), *Intelligent computer tutors: Real world applications* (pp. 179–203). New York: Taylor & Francis.

Imparato, N., & Harari, O. (1993). *Jumping the curve: Innovation and strategic choice in an age of transition.* San Francisco: Jossey-Bass.
Katz, J. E. (1990, October). Caller-ID, privacy, and social processes. *Telecommunications Policy,* pp. 372–410.
Kellogg, W. A., & Thomas, J. C. (1993). Cross-cultural perspectives on Human-Computer Interaction: A report on the CHI 92 workshop. *SIGCHI Bulletin, 25*(2), 40–45.
Kraut, R., Dumais, S., & Koch, S. (1989). Computerization, productivity, and quality of work-life. *CACM, 32*(2), 220–238.
Landauer, T. K. (1995). *The trouble with computers: Usefulness, usability, and productivity.* Cambridge, MA: MIT Press.
Landman, J. (1986). Psychology in a new tempo: Instances and consequences of a cult of celerity. *Computers in Human Behavior, 2,* 287–299.
Peters, T. (1993, October). A paean to self-organization. *Forbes,* pp. 156–157.
Postman, N. (1986). *Amusing ourselves to death: Public discourse in the age of show business.* New York: Viking/Penguin.
Postman, N. (1994, April). Informing ourselves to death. *CTG News,* pp. 3–9.
Rosenschein, J., & Zlotkin, G. (1994). Designing conventions for automated negotiations. *AI Magazine, 15*(3), 29–46.
Stone, A. R. (1991). Will the real body please stand up? Boundary stories about virtual cultures. In M. Benedikt (Ed.), *Cyberspace: First steps* (pp. 81–118). Cambridge, MA: MIT Press.
Stuart, R., & Thomas, J. C. (1991). Virtual reality in education. *Multimedial Review, 2*(2), 17–27.
Thomas, J. (1978). A design-interpretation analysis of natural English with applications to man-computer interaction. *International Journal of Man-Machine Studies, 10,* 651–668.
Thomas, J. (1980). The computer as an active communications medium. *Proceedings of the 18thth Annual Meeting of the Association for Computational Linguistics,* pp. 83–86.
Thomas, J. (1983). Studies in office systems I: The effect of communication medium on person perception. *Office Systems Journal, 1*(2), 75–88.
Walker, J. H., Sproul, L., & Subramani, R. (1994). Using a human face in an interface. *Proceedings of CHI 94* (pp. 85–91). Reading, MA: Addison-Wesley.

Author Index

Page numbers in *italics* denote complete bibliographical references

Subject Index

A

B

C

D

E

L

U